Springer Transactions in Civil and Environmental Engineering

Springer Transactions in Civil and Environmental Engineering (STICEE) publishes the latest developments in Civil and Environmental Engineering. The intent is to cover all the main branches of Civil and Environmental Engineering, both theoretical and applied, including, but not limited to: Structural Mechanics, Steel Structures, Concrete Structures, Reinforced Cement Concrete, Civil Engineering Materials, Soil Mechanics, Ground Improvement, Geotechnical Engineering, Foundation Engineering, Earthquake Engineering, Structural Health and Monitoring, Water Resources Engineering, Engineering Hydrology, Solid Waste Engineering, Environmental Engineering, Wastewater Management, Transportation Engineering, Sustainable Civil Infrastructure, Fluid Mechanics, Pavement Engineering, Soil Dynamics, Rock Mechanics, Timber Engineering, Hazardous Waste Disposal Instrumentation and Monitoring, Construction Management, Civil Engineering Construction, Surveying and GIS Strength of Materials (Mechanics of Materials), Environmental Geotechnics, Concrete Engineering, Timber Structures.

Within the scopes of the series are monographs, professional books, graduate and undergraduate textbooks, edited volumes and handbooks devoted to the above subject areas.

The proposal for each volume is reviewed by the main editor and/or the advisory board. The chapters in each volume are individually reviewed single blind by expert reviewers (at least four reviews per chapter) and the main editor.

Sushree Sunayana · Sudhirkumar V. Barai

Green Concrete

A Sustainable Construction Material

Springer

Sushree Sunayana
Department of Civil Engineering
Malaviya National Institute of Technology
Jaipur (MNIT)
Jaipur, Rajasthan, India

Sudhirkumar V. Barai
Department of Civil Engineering
IIT Kharagpur
Kharagpur, West Bengal, India

ISSN 2363-7633 ISSN 2363-7641 (electronic)
Springer Transactions in Civil and Environmental Engineering
ISBN 978-981-96-6981-3 ISBN 978-981-96-6982-0 (eBook)
https://doi.org/10.1007/978-981-96-6982-0

This Springer imprint is published by the registered company Springer Nature Singapore Pte Ltd.
The registered company address is: 152 Beach Road, #21-01/04 Gateway East, Singapore 189721, Singapore

Dedicated to the Pioneers of sustainable construction, Whose groundbreaking work paves the way for a greener future.

Preface

The construction industry is undergoing a profound transformation driven by the pressing need to adopt sustainable practices and reduce environmental impact. At the forefront of this shift is the innovative recycling of concrete waste to serve as alternative materials for new construction. The production of coarse recycled concrete aggregate (RCA) from construction demolition waste represents a critical advancement in recycling, providing a sustainable solution to the depletion of natural resources. However, incorporating RCA into concrete presents unique challenges, necessitating a thorough understanding of recycling methods, material performance, and the behaviour of the resulting concrete. Traditional concrete production is associated with high carbon emissions due to cement manufacturing. This book delves into using coarse RCA in concrete, particularly when combined with supplementary cementitious materials (SCMs) such as fly ash, slag, and silica fume.

Recycled aggregate concrete (RAC) faces significant challenges. The quality of RCA can vary depending on the source of the demolished concrete, processing methods, crushing techniques, and the presence of impurities, leading to inconsistent properties in the final concrete mix. The attached mortar layer on the surface of coarse RCA creates two interfacial transition zones (ITZ) when used in concrete, which are the weakest phases in the material, adding to its inherent weaknesses. This adhered mortar reduces density and increases porosity, negatively impacting strength, durability, and water absorption. These challenges can be effectively addressed by strategically using SCMs such as fly ash, slag, and silica fume, which partially replace cement and enhance sustainability while improving macro-level properties and microstructural characteristics. This book provides an in-depth examination of these effects, mainly focusing on fly ash due to its availability and beneficial properties. Through case studies and experimental results, the book demonstrates the effectiveness of fly ash as a cement substitute.

Throughout its chapters, this book offers comprehensive discussions on the macro-level properties of RAC, including its fresh and hardened states. This book addresses safety, sustainability, and serviceability concerns by converting waste into valuable resources by incorporating SCMs in RAC. Advanced characterization techniques are used to explore the microstructural properties, revealing the interactions between

SCMs and RCA at a microscopic level. Additionally, the behaviour of structural elements—such as beams, columns, and slabs—constructed with RAC and SCMs is meticulously analysed, providing valuable insights for engineers and researchers. The integration of these materials not only mitigates the drawbacks of RCA but also advances high-performance, environmentally friendly concrete solutions.

This book explores the application of alternative materials like coarse recycled concrete aggregates and SCMs in concrete, focusing on sustainability, strength, safety, and durability. It examines concrete at various scales, from micrometres to metres, covering micro- and structural levels. The book is organized into 11 chapters, divided into four parts. Part I addresses sustainability concerns and green concrete definitions. In contrast, subsequent parts explore the properties and impacts of SCMs on RAC, mix design methods, environmental impact assessments, fresh and hardened concrete properties, microstructural analyses, and structural applications. The final chapters discuss durability, long-term properties, and future trends in sustainable concrete.

This book aims to be an essential resource for civil engineers, materials scientists, researchers, and students involved in developing and applying sustainable construction materials. By providing a thorough understanding of the integration between recycled aggregates and SCMs, it seeks to inspire innovation and promote greener construction practices. The authors hope this book will contribute significantly to developing sustainable concrete solutions.

Jaipur, India Sushree Sunayana
Kharagpur, India Sudhirkumar V. Barai

Acknowledgements

The authors wish to extend their deepest gratitude to the numerous researchers who have significantly contributed to construction demolition waste recycling and reuse. Their extensive knowledge and groundbreaking work have been invaluable in the creation of this book. Our understanding and ability to write on this subject would not have been possible without their pioneering research.

We sincerely thank the Ministry of Human Resource Development, the Government of India, and the Indian Institute of Technology Kharagpur for their critical financial support. This funding has been instrumental in facilitating our research endeavours, enabling us to push the boundaries of our work. The advanced laboratory facilities provided by the structural lab in the Civil Engineering Department and the central research facilities at IIT Kharagpur have been essential to our research, providing the necessary tools and environment to conduct our studies effectively.

We are immensely grateful to Prof. Arghya Deb, Prof. Sushanta Chakraborty, Prof. Nirjhar Dhang of the Civil Engineering Department, and Prof. Debashis Deb of the Mining Department. Their invaluable suggestions, constructive feedback, and expert advice at various stages of this work have been crucial in refining our research and enhancing the overall quality of this book. Their insights have been a guiding light, steering us towards new perspectives, and deeper understanding.

Special thanks are owed to our parents and family members. Their patience, understanding, and encouragement have been the bedrock of support during the long hours of research and writing. Their unwavering belief in our work has been a constant source of motivation and inspiration, driving us to persevere even when faced with challenges.

Dr. Sushree Sunayana would like to express her profound gratitude to Prof. Sudhirkumar V. Barai for his invaluable guidance, constant supervision, and cooperation throughout her Ph.D. programme. His continuous academic support and mentorship have successfully completed her research work. His encouragement and expert advice have greatly influenced the direction and outcomes of this study.

Dr. Sushree Sunayana also extends her deepest gratitude to her husband, Dr. Subha Ghosh. His valuable suggestions, unwavering support, and encouragement have been integral to this journey. His patience, understanding, and positivity have provided a steadfast source of inspiration and strength, enabling her to navigate the complexities of research and writing confidently.

This work represents a collective effort, drawing on many individuals' expertise, support, and inspiration. We extend our heartfelt thanks to everyone who has contributed to this endeavour, whether through direct involvement or through the broader impact of their work. Through such collaborative efforts, we can advance the field of sustainable construction and make meaningful progress towards a more environmentally friendly future.

We are profoundly thankful to the editorial team at Springer for their unwavering support and expert guidance throughout the publication process. Their dedication and meticulous attention to detail have greatly enhanced the quality and coherence of this work.

Finally, we acknowledge the broader community of engineers, materials scientists, and students dedicated to pursuing sustainable construction. Their passion and commitment continue to inspire and drive innovation in this field. We hope this book will serve as a valuable resource for all who promote environmentally friendly concrete and sustainable construction practices.

In conclusion, the journey of writing this book has been enriching and enlightening. The support and contributions of many have made it possible, and we are profoundly grateful. We hope this book not only provides valuable insights and knowledge but also inspires continued research and innovation in the field of sustainable construction materials.

Sushree Sunayana, Ph.D.
Sudhirkumar V. Barai, Ph.D.

Contents

Part I Green Concrete Materials, Mixing, Mix Proportioning and Environmental Impact

1 Introduction 3
1.1 Life-Cycle of a Concrete Structure 3
1.2 Sustainability Challenges of Concrete 4
1.3 Utilization of Waste for Addressing the Sustainability Concerns 7
References 9

2 Production and Characterization of Materials 11
2.1 Production of Coarse Natural Concrete Aggregates 11
2.2 Production and Processing of Coarse Recycled Concrete Aggregates 12
2.3 Surface Treatment Methods of Coarse RCA 15
2.3.1 Reducing Adhered Mortar Content 15
2.3.2 Increasing the Strength of Adhered Mortar Layer 17
2.4 Properties of Coarse RCA and Their Significance 18
2.4.1 Shape and Surface Texture 18
2.4.2 Density and Specific Gravity 19
2.4.3 Porosity and Water Absorption 20
2.4.4 Crushing Strength, Impact Strength, and Abrasion Resistance 21
2.4.5 Elastic Modulus 23
2.4.6 Alkali-Silica Reaction (ASR) 24
2.4.7 Thermal Properties of Aggregate 25
2.5 Properties of Supplementary Cementitious Materials (SCMs) 25
2.5.1 Fly Ash 25
2.5.2 Slag 26
2.5.3 Silica Fume 26

2.6 Summary ... 27
References ... 27

3 Mixing and Mix Proportioning Methods ... 33
3.1 Material Mixing Methods and Their Significance ... 33
3.1.1 Conventional Method ... 33
3.1.2 Two-Stage Mixing Method ... 34
3.1.3 Triple Mixing Method ... 34
3.2 Mix Design Methods ... 35
3.2.1 Conventional Mix Design ... 35
3.2.2 Equivalent Volume Method ... 35
3.2.3 Particle Packing Method ... 36
3.3 Summary ... 43
References ... 44

4 Environmental Impact of Concrete ... 47
4.1 Environmental Impact of RAC ... 47
4.2 Environmental Impact Assessment of RAC Containing SCMs ... 49
4.2.1 Effect of Fly Ash and Coarse RCA ... 49
4.3 Summary ... 57
References ... 58

Part II Properties of Green Concrete

5 Fresh Concrete Properties ... 63
5.1 Workability ... 63
5.1.1 Role of Recycled Aggregates ... 63
5.1.2 Role of Cementitious Materials ... 65
5.2 Fresh Density ... 66
5.3 Air Content ... 67
5.4 Bleeding and Segregation ... 67
5.5 Summary ... 67
References ... 68

6 Hardened Concrete Properties ... 71
6.1 Effect of Coarse RCA on Compressive Strength (CS) and Tensile Strength of Concrete ... 71
6.2 Effect of Coarse RCA on Elastic Modulus of Concrete ... 73
6.3 Effect of Fly Ash on Compressive Strength of Coarse RCA-Based Concrete ... 73
6.4 Effect of Slag and Silica Fume on Compressive Strength of Coarse RCA-Based Concrete ... 78
6.5 Effect of Fly Ash on Tensile Strength of Coarse RCA-Based Concrete ... 79

6.6 Effect of Slag and Silica Fume on Tensile Strength of Coarse RCA-Based Concrete ... 81
6.7 Effect of Fly Ash on Modulus of Elasticity (E) of Coarse RCA-Based Concrete ... 82
6.8 Effect of Slag and Silica Fume on Elastic Modulus of Coarse RCA-Based Concrete ... 82
6.9 Summary ... 82
References ... 84

7 Microstructure Properties ... 87
7.1 Interfacial Transition Zone of RAC-Based Concrete ... 87
7.2 Sample Preparation and Working Principle ... 89
7.2.1 Thermogravimetric Analysis (TGA) ... 89
7.2.2 Scanning Electron Microscope Imaging ... 92
7.2.3 Nanoindentation Study ... 94
7.2.4 Non-destructive Scanning Using X-Ray Microtomography Test ... 95
7.3 Degree of Hydration ... 96
7.3.1 Effects of Curing Age ... 97
7.3.2 Effect of Fly Ash in NAC ... 98
7.3.3 Effect of Aggregate Types ... 101
7.4 Distribution of Hydration Products ... 103
7.4.1 Porosity, CH, and Unhydrated Phase Content (%) ... 106
7.5 Hardness and Modulus of Interface ... 107
7.6 Distribution of Pores in the Specimens in Three Dimension ... 112
7.7 Relation Between Micro and Macro-level Properties ... 117
7.8 Effect of Other Cementitious Materials on the Microstructure of RAC ... 119
7.9 Summary ... 120
References ... 121

Part III Structural Performance Assessment

8 Behaviour of Reinforced Concrete Beams ... 125
8.1 Behaviour of RAC Beams Under Flexure ... 125
8.2 Behaviour of FARAC Beams Under Flexure ... 128
8.2.1 Load-Carrying Capacity ... 128
8.2.2 Crack Propagation and Failure Pattern ... 134
8.2.3 Tension Stiffening of FARAC Beams: A Numerical Investigation ... 135
8.2.4 Reliability Assessment of Serviceability Limit State of Beams ... 141
8.3 Behaviour of Beams Under Shear ... 148
8.3.1 Shear Behaviour of RAC Beams ... 149
8.3.2 Shear Behaviour of FARAC Beams ... 152

8.3.3 Applicability of Existing Design Provisions 164
8.3.4 Reliability of Existing Design Provisions 167
8.4 Summary 177
References 179

9 Behaviour of Reinforced Concrete Slabs Under Punching Shear 183
9.1 Effects of Coarse RCA on Punching Shear Behaviour of Slabs 183
9.2 Effects of Fly Ash and RAC 184
9.2.1 Slab Specimen Details and Loading Set-Up 184
9.2.2 Load-Carrying Capacity 186
9.2.3 Crack Propagation and Failure Pattern 186
9.2.4 Load-Concrete Strain Relation 189
9.3 Applicability of Existing Design Provisions 189
9.4 Reliability of Existing Design Provisions 190
9.4.1 Processing of Database for Calculation of Model Factor (B_{sh}) 191
9.4.2 Reliability Analysis: Results and Discussions 193
9.5 Summary 194
References 195

Part IV Durability and Long-Term Concrete Properties

10 Durability and Long-Term Properties of Concrete with SCMs and Coarse RCA 199
10.1 Permeability 199
10.1.1 Role of Coarse RCA 199
10.1.2 Role of SCMs in RAC 200
10.2 Chloride Attack 200
10.2.1 Role of Coarse RCA 201
10.2.2 Role of SCMs in RAC 201
10.3 Sulphate Attack 202
10.3.1 Role of Coarse RCA 202
10.3.2 Role of SCMs in RAC 203
10.4 Carbonation 203
10.4.1 Role of Coarse RCA 204
10.4.2 Role of SCMs in RAC 204
10.5 Shrinkage 205
10.5.1 Role of Coarse RCA 205
10.5.2 Role of SCMs in RAC 206
10.6 Creep 207
10.6.1 Role of Coarse RCA 208
10.6.2 Role of SCMs in RAC 208
10.7 Summary 208
References 210

11 Recent Advances, Recommendations, and Way Forward 213
11.1 Challenges of RAC and Recent Advances 213
11.2 Way Forward 215
References 216

Part I
Green Concrete Materials, Mixing, Mix Proportioning and Environmental Impact

Chapter 1
Introduction

Preview

This chapter focuses on the sustainability concerns associated with the widely used construction material 'concrete'. The gravity of this issue is discussed using relevant statistics in a global perspective. In this context, green concrete is defined and several methods for enhancing the environmental sustainability of concrete production are explored. The scope of this book is limited to exploring recycling and reusing methods as one of the best possible ways for making sustainable concrete. Potential of alternative materials like coarse recycled concrete aggregates and supplementary cementitious materials is assessed with respect to strength, safety, and sustainability of concrete. We will examine concrete at several scales, ranging from micrometres to metres, encompassing both the microlevel and the level of structural components.

1.1 Life-Cycle of a Concrete Structure

Concrete is the predominant material in building construction, and its utilization has experienced substantial growth in recent decades. This is due to its cost-effectiveness, capacity to be handled by less-skilled workers, and its satisfactory performance in terms of strength and serviceability. The global consumption of concrete is approximately 30 billion tonnes per year (Monteiro et al. 2017). Let us examine the life-cycle of concrete to gain a comprehensive understanding of its environmental effects at various stages. We will also explore how the principles of recycling and reusing can help mitigate the sustainability challenges associated with concrete, which will be covered in the next section.

Figure 1.1 depicts the life-cycle of concrete. In general, there are five steps (Basavaraj and Gettu 2022). The first step is production of raw materials, which are then transported to the concrete batching plant. At the batching plant, various components are mixed in appropriate proportions to produce concrete. After that, concrete

S. Sunayana and S. V. Barai, *Green Concrete*, Springer Transactions in Civil and Environmental Engineering, https://doi.org/10.1007/978-981-96-6982-0_1

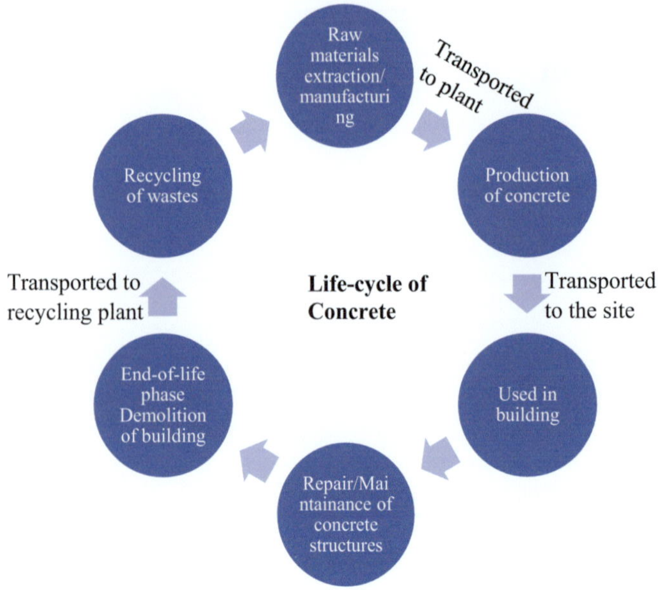

Fig. 1.1 Life-cycle of concrete

is transported and delivered to the location where a structure will be constructed. When properly maintained, the structure fulfils its intended function over its desired service life. The structure needs to be demolished when it becomes unserviceable, even after maintenance, meaning it has reached the limit of its intended functionality or end-of-life. Subsequently, construction and demolition wastes (CDW) are generated. These wastes may either be disposed of in a landfill or transported to recycling facilities for segregation and extraction of raw materials, such as aggregates, after adequate processing for using as alternative concrete constituents. Thus, recycling and reusing of wastes as resources for fresh concrete completes the cycle and concrete becomes circular. The final phase holds great importance within the framework of resource optimization and waste minimization, which are two crucial pillars of sustainability. Achieving sustainability in concrete production has become a global concern. The following section discuss this using some recent data.

1.2 Sustainability Challenges of Concrete

The basic materials necessary to produce concrete include admixture, water, sand, cement, and coarse aggregate. These are the main ingredients that, when mixed in the right amounts, constitute good quality concrete. A standard mix design estimate for M30 grade concrete shows that manufacturing 1cubic metre of concrete requires 1 tonne of coarse aggregate, 0.6 tonne of sand, 0.42 tonne of cement, and 0.2 tonne of water. The fuel and energy consumption involved in extracting, processing, and transporting natural resources such as rock to produce coarse aggregates are substantial. The cement production involves combustion of limestone, resulting in the significant release of carbon dioxide into the environment. Thus, concrete exerts large adverse impact on the environment because of its contribution to resource depletion, CO_2 productions, and waste generation; this calls for quick attention and decisive action. To further visualize the situation, let us have a look at some statistics.

According to the latest data released by the US Geological Survey (USGS) in 2022, global cement production in 2022 reached approximately 4100 million tons. This indicates a decrease of approximately 7% compared to the prior year. In 2022, most of the global cement production, amounting to roughly 2100 million tons (50% of the total), is contributed by developing Asian countries such as China. India accounts for 9% of the total cement production which amounts to 370 million tons in 2022 (Fig. 1.2). Over the past decade, there has been a consistent rising trend in cement production across most nations, except for Egypt, where a significant decline was noted in 2018, followed by a period of relative stability. Although there was a decline in cement production in many nations between 2019 and 2020, the subsequent upward trend indicates a matter of serious concern globally. The environmental burden is magnified by the production of large quantities of cement for the following reasons. The production of cement primarily comprises two processes (Vatopoulos and Tzimas 2012) that majorly contribute to CO_2 emissions. Combustion accounts for 40% and calcination accounts for 60% of the total cement production related emissions. During the calcination process, limestone and clay are subjected to a temperature of about 1450°C to form clinker which then combine with gypsum to produce cement. This calcination process releases huge amount of CO_2. In addition, CO_2 emissions from cement as a percentage of global emissions have been continuously increasing. Currently, cement production attributes towards 10% of the global CO_2 emissions (IEA 2020; Cao et al. 2021). A 1.5% annual increase in CO_2 emissions from cement is observed from 2015 to 2021. To address this issue, the focus has been shifted towards fulfilling the objectives of the 2015 Paris climate change accord, which aims to restrict global warming to a maximum of -1.5 °C. Watari et al. (2022) have emphasized the importance of reducing cement consumption in concrete manufacturing, improving energy efficiency, decreasing the ratio of clinker to cement, carbon capturing and retention, and utilizing other supplementary cementitious materials (SCMs) as binder.

Among these alternatives, the utilization of SCMs can greatly contribute to the decrease in the amount of cement used. This approach can be promptly applied as a solution due to availability of other cementitious materials. The international energy agency's technology roadmap aims to achieve a 24% reduction in global CO_2 emissions from the cement industry by 2050. It is predicted that around 37% of this reduction would be achieved through the use of supplemental cementitious materials (SCMs). SCMs are alternative materials that can partially or completely replace cement. SCMs are classified in to two groups such as self-cementing or pozzolans based on their nature (Paris et al. 2016). An example of a self-cementing material is blast furnace slag, which is a residual product derived from the iron

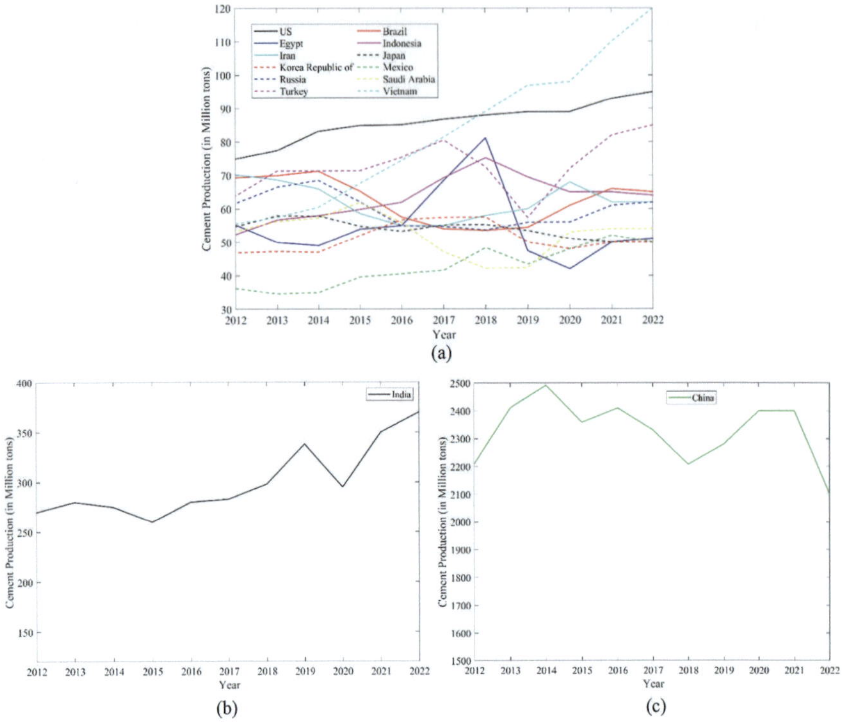

Fig. 1.2 Cement production trend of top fourteen countries producing cement in the last ten years (*Source* USGS 2022)

and steel industry. Slag possesses cementitious property similar to cement (Neville 2011). Pozzolans can be categorized into two types like natural pozzolans such as diatomaceous earth, shale, etc., and artificial pozzolans. Artificial or man-made pozzolans primarily consist of industrial by-products (Malhotra and Mehta 1996) such as fly ash derived from coal-based power plants, silica fume obtained from silicon or ferrosilicon industry, rice husk ash resulting from the combustion of grain husk (Gursel et al. 2016) and sugarcane ash produced through the incineration of sugarcane baggage (Rukzon and Chindaprasirt 2012). This book will address the potential application of supplementary cementitious materials (SCMs), for example fly ash, slag, and silica fume, in recycled coarse aggregate-based concrete.

In addition to cement, which is responsible for a greater CO_2 footprint, aggregates are another primary material that raises sustainability concerns in concrete. Aggregates make up over 75% of the overall volume of concrete, primarily sourced from non-renewable resources such as rock. A report released by the Freedonia Group in 2019, as cited by Wang et al. (2021), indicates that there is a consistent global increase in aggregates consumption of approximately 5% every five years. Furthermore, the report predicts that the demand for aggregates will double in the next two to three decades, mostly driven by large infrastructural development and subsequent

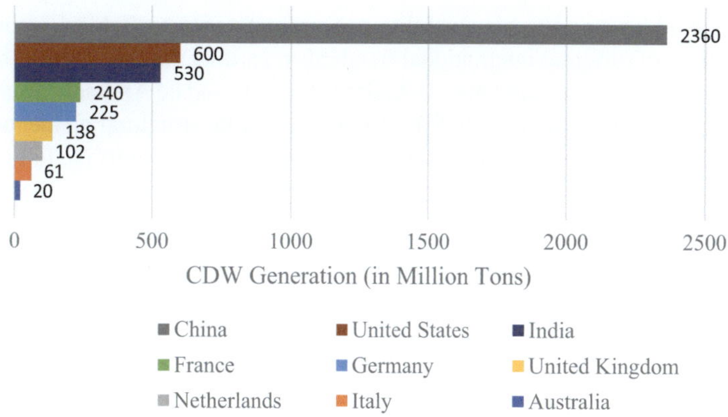

Fig. 1.3 Construction demolition wastes generation by countries (Wang et al. 2021)

rise in concrete construction. The Asian countries, particularly China and India, are supposed to have the highest demand for aggregates, followed by Eastern Europe, Africa, and Middle Eastern countries. This is mostly owing to a significant rise in economic activity. The demand for aggregates in other developed countries like USA, Canada, Japan, Western Europe, and Australia is projected to decrease due to modern construction methods and well-established infrastructures. This growing demand for aggregates points towards an urgent need for alternate source of aggregate, which brings us to the concept of waste recycling and reusing. Figure 1.3 illustrates the production of construction and demolition wastes in various countries for the year 2018, based on data sourced from Wang et al. (2021). China is the leading producer of construction demolition wastes followed by USA and India. The European countries contribute less to CDW as compared to other regions. However, about 35% of the wastes generated is disposed of in landfill, whereas a very smaller percentage is recycled. The recycling rate of CDW in Europe and the USA is approximately 30% (Böhmer et al. 2008). This percentage is even lower in countries like China that is about 5% (Huang et al. 2018) and 9% in Portugal (Coelho and de Brito 2013). Therefore, there is a potential in these CDW to produce coarse recycled aggregates which can be used as a substitute for the natural aggregates. Application of this recycled concrete aggregates (RCA) facilitates resource efficiency and waste minimization.

1.3 Utilization of Waste for Addressing the Sustainability Concerns

In the context of lowering the environmental burden of concrete, it has become imperative to minimize the cement content as well as exploitation of non-renewable resources, such as rocks, for coarse aggregate production. Green concrete made with recycled concrete waste and supplimentary cementitious materials is a promising solution for addressing the sustainability concerns of concrete. Table 1.1 compares the guidelines of different international standards regarding the employment of recycled coarse aggregates in concrete (de Andrade Salgado and de Andrade Silva 2022). Germany permits the use of up to 90% RCA in concrete structural elements, if the water absorption of the aggregates is less than 10%. RILEM permits the full replacement of natural aggregates with the recycled ones. In structural applications, other standards impose a restriction on maximum replacement percentage of up to 20 or 30%. The existence of an adhering mortar layer limits the structural applications of recycled aggregates since it increases water absorption, porosity, and the establishment of two interfacial transition zones. The use of any new materials in construction is limited by the condition that they must meet the performance standards set by existing materials.

Several studies (Kou et al. 2007; Ann et al. 2008; Somna et al. 2012; Ahmad 2013; Sunayana and Barai 2017) have examined the effects of supplementary cementitious materials (SCMs) on concrete made with recycled aggregate. The use of SCMs has been found to reduce the environmental implications of RAC (Marinković et al. 2017; Göswein et al. 2018). They also reported a reduction of environmental impacts of RAC due to the presence of SCMs. Thus, recycled aggregates, when combined with supplementary cementitious materials, can serve as alternative constituents for concrete, if the concrete meets the performance criteria for compressive strength,

Table 1.1 International standard provisions for use of recycled concrete aggregates

Country	Codes	Concrete use	% Replacement allowed	Dry density	Water absorption %	Contaminants non-minerals %
Brazil	NBR 15116	structural	20	–	≤7	≤1
Portugal	LNEC E471	C35/45	25	≥2200	≤7	<0.2
		C40/50	20			<0.5
Germany	DIN 4226-100; 12620	Structural	90	>2000	≤10	≤0.2
			70		≤15	≤0.5
Netherlands	NEN 5905	C45/55	20	>2100	–	–
UK	BS 8500-2	C20/25 C40/50	20	–	–	<0.5
Japan	JIS A 5021	C45/55	–	≥2500	≤3	≤1
China	DG/TJ/ 07/ 008	–	–	–	<10	≤1
Australia	HB 155:2002	C40	30	≥2100	≤6	<1
		C25	100	≥1800	≤8	<2
RILEM		C50/50	100	≥2000	≤10	–
		C16/20		≥1500	≤20	

modulus of elasticity, tensile strength, serviceability, and load carrying capacity of structural components.

This book focuses on the behaviour of recycled aggregate concrete incorporating cementitious materials as cement substitute. Specifically, the impact of fly ash on concrete made entirely from coarse recycled concrete aggregate is explored. To establish this as an alternative concrete constituent, the concrete performance is discussed at material, microstructural as well as structural components levels in the following chapters.

References

ABNT NBR 15116 (2021) Agregados Reciclados de Resíduos Sólidos Da Construção Civil—Utilização Em Pavimentação e Preparo de Concreto Sem Função Estrutural-Requisitos. ABNT, Rio de Janeiro, Brazil

Ahmed SFU (2013) Properties of concrete containing construction demolition wastes and fly ash. J Mater Civ Eng 25(12):1864–1870

Ann K, Moon H, Kim Y, Ryou J (2008) Durability of recycled aggregate concrete using pozzolanic materials. Waste Manage 28(6):993–999

Basavaraj AS, Gettu R (2022) Life cycle assessment as a tool in sustainability assessment of concrete systems: why and how? Indian Concrete J 96(4):1–20

Böhmer S et al (2008) Aggregates case study: final report referring to contract 150787-2007 F1SC-AT

BSI, BS 8500-2 (2015) Concrete—complementary British standard to BS EN 206—part 2: specification for constituent materials and concrete, BSI, London, UK, p 48

Cao Z, Masanet E, Tiwari A, Akolawala S (2021) Decarbonizing concrete: deep decarbonization pathways for the cement and concrete cycle in the United States, India, and China (ClimateWorks Foundation)

Coelho A, De Brito J (2013) Economic viability analysis of a construction and demolition waste recycling plant in Portugal—part I: location, materials, technology and economic analysis. J Clean Prod 39:338–352

de Andrade Salgado F, de Andrade Silva F (2022) Recycled aggregates from construction and demolition waste towards an application on structural concrete: a review. J Build Eng 52:104452

DIN D 4226-100 (2002) Aggregates for mortar and concrete—part 100: recycled aggregates. Deutsches Institut Fur Normung

DIN EN 12620 (2003) Aggregates for concrete. Deutsches Institut Fur Normung

Göswein V, Gonçalves AB, Silvestre JD, Freire F, Habert G, Kurda R (2018) Transportation matters–does it? GIS-based comparative environmental assessment of concrete mixes with cement, fly ash, natural and recycled aggregates. Resour Conserv Recycl 137:1–10

Gursel AP, Maryman H, Ostertag C (2016) A life-cycle approach to environmental, mechanical, and durability properties of "green" concrete mixes with rice husk ash. J Clean Prod 112:823–836

H.B. CSIRO, 155 (2002) Guide to the use of recycled concrete and masonry materials. Commonwealth Scientific and Industrial Research Organisation (CSIRO), Australia, p 80

Huang B, Wang X, Kua H, Geng Y, Bleischwitz R, Ren J (2018) Construction and demolition waste management in China through the 3R principle. Resour Conserv Recycl 129:36–44

International Energy Agency (2020) Energy technology perspectives 2020 (International Energy Agency)

JIS A 5021 (2024) Japan society of civil engineers. New standards on recycled aggregates and molten-slag in Japan

Kou SC, Poon CS, Chan D (2007) Influence of fly ash as cement replacement on the properties of recycled aggregate concrete. J Mater Civ Eng 19(9):709–717
LNEC, E 471-2009 (2009) Guia para utilizaç˜ao de agregados reciclados grossos em bet˜oes de ligantes hidr´aulicos, Laborat´orio Nacional de Engenharia Civil. Minist´erio das Obras Públicas, Transportes e Comunicaç˜oes, Lisboa, PRT, p 8
Malhotra V, Mehta PK (1996) Pozzolanic and cementitious materials. Gordon and Breach Publishers
Marinković S, Dragaš J, Ignjatović I, Tošić N (2017) Environmental assessment of green concretes for structural use. J Clean Prod 154:633–649
Monteiro PJ, Miller SA, Horvath A (2017) Towards sustainable concrete. Nat Mater 16(7):698–699
Neville A (2011) Properties of concrete. Prentice Hall
Paris JM, Roessler JG, Ferraro CC, DeFord HD, Townsend TG (2016) A review of waste products utilized as supplements to Portland cement in concrete. J Clean Prod 121:1–18
RILEM TC 121 (1994) DRG guidance for demolition and reuse of concrete and masonry—specifications for concrete with recycled aggregates, vol 27, pp 557–559
Royal Netherlands Standardization Institute (1997) NEN 5905—aggregates for concrete—materials with a density of at least 2000 Kg/m^3
Rukzon S, Chindaprasirt P (2012) Utilization of bagasse ash in high-strength concrete. Mater Des 34:45–50
Somna R, Jaturapitakkul C, Chalee W, Rattanachu P (2012) Effect of the water to binder ratio and ground fly ash on properties of recycled aggregate concrete. J Mater Civ Eng 24(1):16–22
Sunayana S, Barai SV (2017) Recycled aggregate concrete incorporating fly ash: comparative study on particle packing and conventional method. Constr Build Mater 156:376–386
Technology Roadmap (2018) Low-carbon transition in the cement industry. IEA, Paris, France
The Freedonia Group (2019) Global construction aggregates reports
USGS (2022) USGS mineral program cement report. United States Geological Survey, USA
Vatopoulos K, Tzimas E (2012) Assessment of CO_2 capture technologies in cement manufacturing process. J Clean Prod 32:251–261
Wang B, Yan L, Fu Q, Kasal B (2021) A comprehensive review on recycled aggregate and recycled aggregate concrete. Resour Conserv Recycl 171:105565
Watari T, Cao Z, Hata S, Nansai K (2022) Efficient use of cement and concrete to reduce reliance on supply-side technologies for net-zero emissions. Nat Commun 13(1):4158

Chapter 2
Production and Characterization of Materials

Preview

In the previous chapter, we have discussed the pressing need for a durable and green concrete. Use of recycled concrete aggregates as coarse aggregate replacement and other SCMs as partial cement substitute can become potential alternative constituents in concrete. Aggregates constitute 75–80% of the volume of concrete. In this chapter, we will mainly discuss the production of coarse recycled concrete aggregates which are obtained by recycling, sorting, and processing of the construction demolition wastes at the end-of-life of a concrete structure. Thus, waste collection, segregation, and processing methods to prepare them for use as coarse aggregates in concrete are discussed. While aggregates are mostly inert materials which do not participate actively in chemical reactions, their qualities can still impact the fresh and hardened properties of concrete. Thus, it is crucial to understand the properties of recycled aggregates in contrast to natural aggregates. When we fully substitute recycled coarse aggregates for natural coarse aggregates, it is interesting to see how the characteristics of the coarse aggregates impact the performance of concrete at macro-level. This chapter also discusses the properties of commonly available SCMs, like fly ash, slag, and silica fume, which can be utilized to positively modify the performance of RAC.

2.1 Production of Coarse Natural Concrete Aggregates

The coarse natural concrete aggregates, referred to as NCA (most preferably crushed stones), are sourced from mines or rock deposits through excavation, drilling, or blasting techniques (Langer 2016). Explosives can be used to break the rock into rubble, rendering it suitable for subsequent crushing. Excavators might also be used to dig out the rock. Then the gravels or rock rubbles are collected, loaded onto lorries, and transported to the processing plants. Fuel and energy consumptions are associated

S. Sunayana and S. V. Barai, *Green Concrete*, Springer Transactions in Civil and Environmental Engineering, https://doi.org/10.1007/978-981-96-6982-0_2

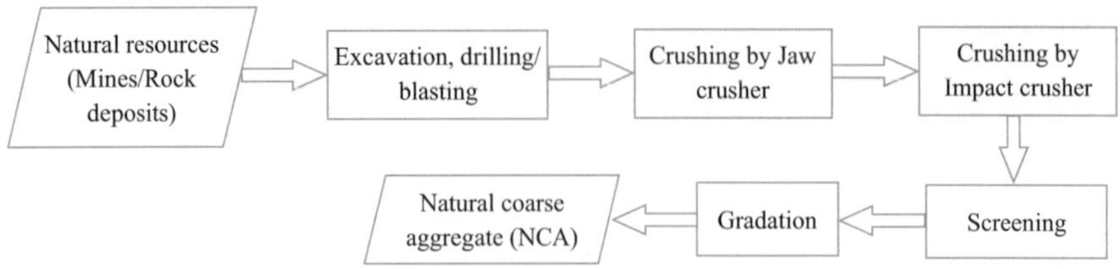

Fig. 2.1 Production processes of natural coarse aggregates

with these excavation and transportation processes. In the processing plants, the rock rubbles are crushed using jaw crushers and impact crushers to break them further down into smaller sizes. After the crushing process, screening of materials is carried out by passing over a screening system followed by sorting into different sizes, with a greater emphasis on producing aggregates of 20 mm down size (which are mostly used in building construction), including other intermediate size ranges. These processes of coarse aggregate extractions are also discussed elsewhere (Hossain et al. 2016). Detailed quantitative impact of these processes are assessed in Chap. 4 on life-cycle analysis. Figure 2.1 depicts the natural coarse aggregate processing steps.

2.2 Production and Processing of Coarse Recycled Concrete Aggregates

Prior to the actual demolition, a thorough examination of the conditions of existing buildings is necessary. This will assist in selecting the suitable approach between complete demolition and selective demolition. For total demolition of any old structures, engineering (mechanical) methods such as the use of wrecking ball, abrasion, explosive or non-explosive, pressure jetting, and thermal processes can be employed; the specific details of these methods can be referred from another source (Rao et al. 2019). The wrecking ball technique involves the dismantling of a structure through the forceful impact of a swinging steel ball weighing around 0.5–1 tonne. Cranes with high capacity and skilled labours are required for better execution. The abrasion method uses hard materials like diamond discs, strings, or carborundum to effectively penetrate concrete (Coelho and de Brito 2013). While this approach is costly, it may be implemented for the partial demolition of structures. Dynamite and other explosives can be utilized for demolition purposes. Alternatively, the non-explosive method involves the expansion of slurry that is injected into special perforations that are positioned along the intended concrete separation line. This process results in the formation of fractures, which in turn effectively demolish the structure (Rao et al. 2019). The demolition method utilizing pressure jets involves the use of highly pressurized water from small jet heads to dislodge aggregates by washing away the soft mortar (Hilmerssion 1998). Although the pressure jetting approach has advantages such as reduced dust and induced vibrations, it is ineffective for cutting through reinforcements. The thermal method of demolition is employed for special structures or inaccessible locations, where steel reinforcements are subjected to extremely high temperature (around 1600 °C) to remove them from concrete or surface concrete is fractured by direct heat application (Coelho and de Brito 2013).

After the demolition process, construction demolition wastes are transported to a recycling plant. The first step of the recycling process involves removal of other wastes which are categorized as follows: (a) materials with high marketable values such as glass, decorative tiles, door, and window units, (b) hazardous materials, (c) woods, plastics, plaster, small bricks, or hardened cement paste. The process also involves the elimination of surface elements that have been contaminated by chemicals, as described by Hansen (1985), Coelho and de Brito (2013), and Silva et al. (2017). The choice between stationary or mobile recycling plants for proper recycling facilities is influenced by factors such as the degree of contamination, the technology available, and the desired quality of recycled aggregates for their intended future applications (Silva et al. 2017). Stationary plants are equipped with large primary crusher with high production capacity, as well as secondary or tertiary crushers and sieving devices to enable increased output capacity. These plants can produce better quality and well-graded RCA with high efficiency. However, the distance for transporting CDW to the fixed recycling facility may vary depending on the demolition site's location. On the other hand, the mobile recycling plant mainly consists of one primary crushing unit and its mobility allows for easy movement to different demolition sites, facilitating lesser transportation distance. Nevertheless, the operation of this plant may lead to increased levels of dust and noise in the vicinity, while also producing inferior quality RCA. A typical mobile crusher is shown in Fig. 2.2.

The key steps involved in any recycling procedure are (a) crushing, (b) sorting, and (c) processing. The schematic diagram in Fig. 2.3 illustrates a standard recycling process, which is based on the works of Hansen (1985) and Wang et al. (2021). The segregation of other wastes or contaminants (such as pieces of wood, plastic, glass, etc.) can be carried out either at the beginning or following the primary and secondary crushing stage. Magnetic separators positioned strategically along the conveyor belts can be utilized to eliminate different types of ferromagnetic metal

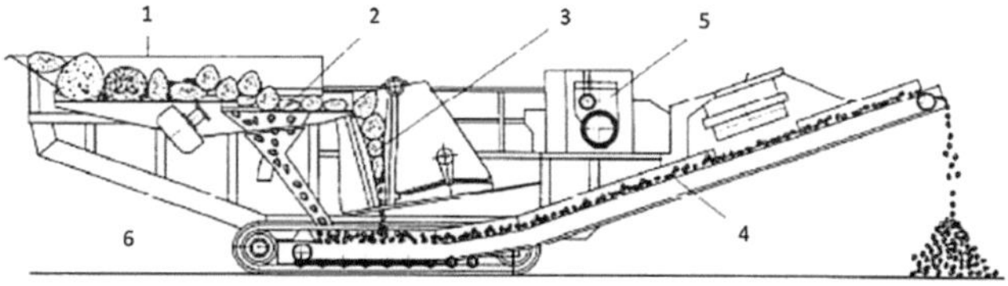

(1. Feeding hopper 2. Oscillating conveyor 3. Jaw crusher 4. Discharging transport belt 5. Diesel engine as power unit 6. Wheels for mobility)

Fig. 2.2 Mobile crusher (adopted from Kumbhar et al. 2013)

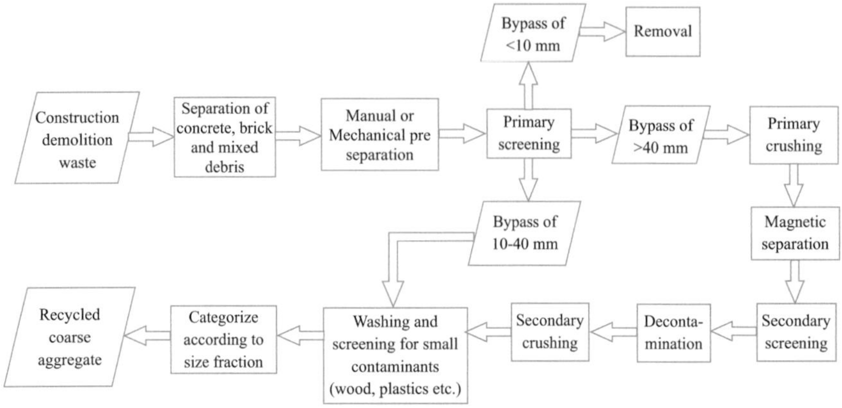

Fig. 2.3 Production procedure of coarse RCA from construction demolition waste

waste and steel reinforcements. Non-ferrous materials like Aluminium, Copper, etc., can be effectively eliminated by installing an eddy current separator device at the end of the conveyor belt, as suggested by Wang et al. (2021). Additional contaminants like dirt, gypsum, and plasters can be removed from the crushed aggregates after the secondary crushing process. The crushing stage comprises utilization of jaw crusher, impact crusher, and gyratory crusher (Silva et al. 2017). Jaw crushers are suitable to break large reinforced concrete pieces into smaller workable size concrete pieces which then passes through the impact crushers. In the jaw crusher, the demolition wastes blocks are passed through two plates one fixed and one oscillating and the oscillatory motion of the later helps in crushing. Impact crushers employ the principle of striking the debris together using a high-speed rotating impact to break a significant portion of larger sized concrete blocks and generates more fines. Gyratory crushers, as the name suggests, may break down smaller concrete wastes by using a gyratory motion of wheels. They also use less energy to produce the desired particle size of aggregates, resulting in fewer fines.

Various recycling methods are available for processing C&D wastes, including ordinary recycling, mechanical grinding, or thermal treatment (de Andrade Salgado and de Andrade Silva 2022). The ordinary recycling method involves following four processes: (a) primary crushing using jaw crusher, (b) removal of iron wastes using magnetic separator, (c) secondary crushing for particles greater than 20 mm, and (d) sieving to achieve the desired particle size gradation of aggregates. In addition to the jaw crusher, the mechanical method employs impact crushers to break the concrete wastes. This process can result in the production of better-quality aggregates by reducing the amount of adhered mortar that sticks to the particles through friction. Thermal treatment method can be followed in the recycling process to further weaken the attached mortar on aggregates. In the heating and rubbing method described by Shima et al. (2005), concrete wastes are crushed to a size of 50 mm and then subjected to a temperature of 300 °C to dehydrate and remove the old mortar layer sticking to the aggregates surface. The industrial manufacturing method of RCA also involves an

Fig. 2.4 Recycling plant **a** conveyer belt, **b** coarse RCA of different size fractions produced after recycling

advanced dry recovery method and an additional heating air (700 °C) classification system to dry the aggregates and eliminate impurities (Gebremariam et al. 2020; Ibrahim et al. 2023). Water sprinklers and suction bags are employed to reduce dust emission during this recycling process in the plants. After undergoing the process of crushing and removing the contaminants, aggregates of different size fractions are sorted and stored in open-air piles.

A recycling plant located in New Delhi (India) is shown in Fig. 2.4. This figure depicts the construction demolition wastes as well as the coarse RCA finally produced after proper recycling and sorting.

2.3 Surface Treatment Methods of Coarse RCA

The coarse recycled concrete aggregates (RCA) obtained in the above-mentioned method is characterized by the presence of an adhering mortar layer on its surface. Due to the presence of microcracks and higher porosity, this attached mortar layer causes inferiorities in the RCA. Hence, there are multiple surface treatment techniques that can be employed to either decrease the amount of adhered mortar or to enhance its strength (Ibrahim et al. 2023). Some of these techniques are discussed below.

2.3.1 Reducing Adhered Mortar Content

The removal of adhered mortar from the surface of recycled concrete aggregate (RCA) can be accomplished via (a) chemical, (b) mechanical, and (c) thermal techniques. In the chemical method of surface treatment (Raman and Ramaswamy 2021), acid solutions such as HCl, or H_2SO_4, with a concentration of 0.1 M, are utilized to presoak the RCA for 24 h. Subsequently, the RCA was subjected to Los Angeles abrasion. The use of these acid solutions is found to improve the specific gravity and reduce the water absorption of aggregates. H_2SO_4 is found to be more effective than HCl. The mechanical method for reducing adhered mortar involves rotating the RCA in a modified concrete mixer at a speed of 5 revolutions per minute for a duration of 5 h (Dimitriou et al. 2018). In this technique, water is added to minimize the dust emission. Adhered mortar reduces by about 9%, resulting in the production of round-shaped aggregate. This method also improves density of aggregates, lower water absorption, and higher workability of concrete.

In the thermal technique (Al-Bayati et al. 2016), coarse RCA is subjected to different temperatures (250 °C, 350 °C, 500 °C, and 750 °C). The optimized temperature with respect to higher specific gravity and lower water absorption is found to be 350 °C. Higher temperature may be beneficial for removing adhered mortar, but it also leads to increased stresses and microcracking on the surface of the aggregate. In the thermal method studied by Pandurangan et al. (2016), the RCA is heated upto 500 °C for 2 h, followed by cooling in water. The results showed an increase in the compressive strength of concrete by 3 MPa as compared to the concrete with untreated recycled aggregates. Furthermore, a combination of chemical and thermal methods known as chemico-thermal (Santha kumar et al. 2019) or thermochemical (Santha kumar and Minocha 2018) was investigated for the removal of adhered mortar. As the name suggests, the chemico-thermal method involves an initial chemical reaction followed by a thermal process. Initially, RCA is soaked in CH_3COOH solution (of molarities 0.5, 1, 5, and 10) for a duration of 24 h followed by heating at 400 °C for 40 min. In addition to improving the specific gravity and reducing water absorption, this treatment method also leads to an increase in the compressive and flexural strength of concrete. The strength gain was found to be the highest for RCA immersed in solution of concentration 1 M, and this method outperformed other methods in terms of performance. On the other hand, the thermochemical method involves subjecting the RCA to a heat treatment of no more than 600 °C for a duration of 2 h followed by immersion in a HCl solution for 24 h. This procedure allows for a removal of around 20% of the adhering mortar, resulting in an enhanced compressive strength of the concrete.

Though these methods discussed above look promising in terms of quality improvement of RCA, the associated environmental or cost-effectiveness concerns cannot be ignored. Heat treatment methods are not only energy intensive but also emit CO_2 in the process (Al-Bayati et al. 2016). The chemical methods involve prolonged soaking of RCA in chemicals can be expensive (Kim et al. 2017). As coarse RCA-based concrete is intended to achieve greening of the conventional concrete, a balance between the performance requirements and other sustainability parameters need to be established.

2.3.2 *Increasing the Strength of Adhered Mortar Layer*

The strength of adhered mortar layer can be enhanced by filling its microcracks and pores through deposition or coating with denser materials. This can be achieved through carbonation, bio-deposition, cement, or polymeric coating and heat treatment (Ibrahim et al. 2023). According to Li et al. (2019), accelerated carbonation reaction occurs when RCA are subjected to 20% concentration of CO_2 gas at a temperature of around 25 °C and a humidity level of 70% for 3 days. During this process, carbonates fill the micropores and densify the adhered mortar layers. In a similar manner, sodium silicate can be used to treat RCA by reacting with CH to produce additional calcium silicate hydrate or C-S-H (Alqarni et al. 2021). This reaction occurs when the RCA is subjected to oven drying at 100 °C for 24 h and subsequently mixed with a sodium silicate solution of varying concentrations (20, 30, 40, and 50%) for a duration of 30 min. This treatment process helps in reducing the abrasion value and water absorption of RCA as well as increasing the compressive strength of concrete. To strengthen the adhered mortar layer, RCA may be coated with different materials like polymer emulsion (Spaeth and Tugguer 2013), cement (Zhang et al. 2021), pozzolanic materials (Sasanipour and Aslani 2020), or bio-cement (Mistri et al. 2021). Polymer emulsions containing silane or siloxane compounds are coated on RCA for a duration of five minutes and thereafter allowed to air dry to reduce water absorption. Similarly, to enhance sulphate attack on concrete, RCA may be coated with a sulpho-aluminate cement slurry with a higher w/c ratio of 0.8 (Zhang et al. 2021). Pozzolanic materials, such as silica fume slurry, can be coated on RCA followed by over-drying at 45 °C for three days. The coating of pozzolanic materials can improve the chloride resistance of concrete. A bacterial solution can be sprayed on the dry RCA surface for a duration of three hours followed by cement coating to improve the specific gravity of RCA and compressive strength of concrete (Mistri et al. 2021). Wu et al. (2020) utilized the bio-deposition technology to enhance the strength of RCA. In this experiment, RCA is submerged in a solution containing H4 bacillus bacteria for a duration of 20 days. This treatment method can reduce water absorption, porosity, and consequently increase the compressive strength of concrete. This bacteria culture solution can also be sprayed on RCA surface for similar positive effects of strengthening the old mortar.

While aggregates are mostly inert materials which do not participate actively in chemical reactions, their qualities can still impact the fresh and hardened properties of concrete. Thus, it is crucial to understand the properties of coarse RCA and how recycled aggregates differ from natural aggregates in this regard.

2.4 Properties of Coarse RCA and Their Significance

The properties of coarse aggregate that essentially affect the characteristics of concrete include shape, surface texture, porosity, water absorption, density, crushing strength, and elastic modulus among others (Mehta and Monteiro 2006). These properties of aggregates are shaped by the nature of parent rock as well as the crushing and production methods. As discussed in the previous section, RCA is produced from recycling of construction demolition wastes using different stages like demolition, sorting, screening, and crushing. The parent aggregates may also have experienced degradation during the service life of the original structure. The factors influencing the characteristics of RCA are complex and can vary from those of natural coarse aggregates. Understanding these properties is essential for effectively utilizing coarse RCAs in concrete. The following section discusses the properties of coarse RCA.

2.4.1 *Shape and Surface Texture*

The shape and surface texture of coarse aggregates are mainly influenced by their prior exposure to external forces, as well as the methods used for crushing and processing. The shape of aggregates affects its packing potential in the concrete mix, workability as well as mechanical properties of concrete. The surface texture of aggregate has a significant impact on its bonding with the cement paste. Typically, a coarse RCA particle comprises of 65–70% part as parent aggregate and the remaining part as the attached mortar (Zhang et al. 2015). In the process of obtaining recycled aggregates through crushing, the shape and surface texture are influenced by the parent rock, as well as the crushing methods, levels, and reduction ratio. Crushing methods and levels of crushing are explained in the previous section.

Few studies (Ravindrarajah and Tam 1985; Bairagi et al. 1993) have reported that coarse RCAs mostly exhibit angular shape due to the crushing methods. The presence of adhering mortar gives these aggregates a rough and porous surface texture. The shape and flakiness index of coarse RCA have been documented in several literatures previously. Both RCA and NCA exhibited a shape coefficient of 0.363 for aggregates with size ranging from 10 to 20 mm (Gómez-Soberón 2002). Additionally, the shape index was found to be 28 and 25% (Etxeberria et al. 2007). The flakiness index of coarse RCA (4–12 mm size) was lower (9%) compared to that of NCA (25%) due to its angular shape with many cracking faces (González-Fonteboa and Martinez-Abella 2008). Sunayana and Barai (2017) used the Indian Standard IS:2386 (part-3) and reported equivalent elongation and flakiness index (shown in Table 2.1). This indicated relatively rounded shape of RCA. As per Indian provisions, deviations from equidimensional shape result in elongated and flaky aggregate particles. The elongation index, as measured by a length gauge, is defined as the total percentage of aggregates that have a length greater than 1.8 times the mean sieve size. On the other hand, the flakiness index, as measured by thickness gauge, refers to the percentage of

Table 2.1 Physical properties of coarse aggregates

Aggregate type	Size range (mm)	Specific gravity	Bulk density		Water absorption (%)	Flakiness index	Elongation index
			Loose (Kg/m^3)	Compacted (Kg/m^3)			
NCA	20–12.5	2.9	1450.00	1599.27	1.59	19.14	49.42
	12.5–10	2.88	1440.00	1583.16	1.67		
	10–6.3	2.8	1425.00	1549.47	1.06		
RCA	20–12.5	2.54	1303.86	1374.00	3.63	15.17	18.24
	12.5–10	2.53	1230.87	1315.08	3.4		
	10–6.3	2.43	1191.60	1271.60	4.2		
	6.3–4.75	2.29	1146.50	1288.00	5.09		

particles with a thickness that is less than 0.6 times the least lateral dimension. The shape of aggregate is mostly affected by the crushing methods, number of crushing stages, and type of crushers employed to produce RCA from larger pieces of CDW (Silva et al. 2014).

As previously discussed, the recycling of RCA involves two stages of crushing, i.e. primary and secondary. Jaw crushers are used for primary crushing, which leads to the production of flat and more sharp or angular aggregate particles. According to Pedro et al. (2015), the process of secondary crushing using either an impact crusher or cone crusher produces less sharp, less flaky, and relatively rounded particles. The shape index of the RCA was slightly lower than that of the NCA as a result of the tertiary crushing procedure (de Brito et al. 2016). Due to the attached old mortar layer on the surface of RCA, it is likely that the RCA will have relatively rounder shape as compared to coarse NCA, even if both undergo a similar crushing process.

2.4.2 Density and Specific Gravity

Bulk density of coarse aggregates is defined as the mass of aggregates contained in a unit volume. The volume is occupied by both the aggregate particles and voids in between them. Bulk density measures mass per unit volume including voids. The bulk density of typical natural aggregates ranges from 1300 to 1700 kg/m^3. The bulk density value for coarse RCA was reported to be between 1140 and 1400 kg/m^3 (Topçu and Guncan 1995; Rao et al. 2011; Ferreira et al. 2011). The density of coarse RCA with a size of 10–25 mm was found to be 10% lower than that of coarse NCA (González-Fonteboa and Martinez-Abella 2008). The bulk density of coarse RCA is lower than that of NCA due to the presence of porous and less dense adhered mortar layer (de Brito et al. 2016). The density of aggregates is directly influenced by the quantity of the old mortar, which can be controlled by adjusting the number of crushing stages. According to Nagataki et al. (2004), utilizing multiple crushing methods is more effective in removing adhered mortar from aggregates, resulting in higher density compared to using a single crushing method. Strength of parent concrete may also have an influence on the density of coarse RCA. For example, RCA with three different strengths (60.7, 49.0, and 28.3 MPa) exhibited saturated surface-dry densities of 2420, 2410, and 2370 kg/m^3, respectively (Gokce et al. 2011).

Surface-saturated dry (SSD) density is a measurement of the density of aggregate particles when the internal pores are completely filled with water but the surface is dry. SSD density is used in mix designs to accurately account for the moisture content of aggregates, ensuring the correct water-to-cement (w/c) ratio. Oven-dry (OD) density is the density of aggregate particles after they have been dried in an oven at a specified temperature (usually 105 °C) until they reach a constant mass. This condition guarantees that all of the aggregate's moisture, including pore moisture inside and on the surface, has been eliminated. The SSD and OD density of laboratory crushed concrete aggregates (size < 25.4 mm) were measured as 2526 kg/m^3 and 2448 kg/m^3, respectively. In contrast, the natural coarse aggregate had marginally higher values at 2573 and 2550 kg/m^3 (Gomes and de Brito 2009). The similar trend was also observed for CDW collected from a mobile recycling plant, where the larger coarse RCA in a size range of 4–25.4 mm had SSD density of 2440 kg/m^3 and an OD density of 2300 kg/m^3 (Ferreira et al. 2011). In comparison, the coarse NCA had an SSD density of 2640 kg/m^3 and OD density of 2600 kg/m^3, respectively. Sunayana and Barai (2017) used the Indian Standard IS:2386 (part-3) to quantify the physical properties of aggregates as outlined in Table 2.1. RCA had a lower specific gravity and a higher water absorption in comparison with NCA. This observation can be related to the presence of adhered mortar content on RCA (Otsuki et al. 2003; Tangchirapat et al. 2010).

2.4.3 Porosity and Water Absorption

Typically, the water absorption of NCA is within a range of 0.5–1.5% (Silva et al. 2014), which is far too low to be taken into account in the mix design. However, coarse RCA has a very high water absorption capacity that ranges from 3.1% to 7% (Topçu and Guncan 1995; Santos et al. 2002; Rao et al. 2011; Tam et al. 2007; Poon et al. 2007; Berndt 2009). This can be attributed to the porous and weak old mortar layer. The water absorption is influenced by factors such as recycling methods, quality of parent concrete, and size of coarse RCA. The secondary crushing method helps in reduction of adhered mortar content which is one of the main reasons for reducing porosity leading to lower water absorption (Silva et al. 2014). The observations regarding the effects of parent concrete quality are not conclusive. Several researches (Gokce et al. 2011; Andreu and Miren 2014; Poon et al. 2004) have showed that water absorption of coarse RCA decreases as the parent concrete strength increases. Another study (Katz 2004) reported no significant relation between these two parameters. Water absorption varies inversely with the size of aggregate particles; finer the size greater is the water absorption due to greater accumulation of adhered mortar (Katz 2003).

The water absorption of coarse RCA is also influenced by the time of soaking, as observed by Vieira et al. (2011). This information can be used to determine the appropriate amount of water to be added during the process of concrete mix proportioning. Based on their study, 80% of water gets absorbed within the initial five minutes of immersion of aggregates in water. After 30 min, only an additional 4% water is absorbed. Sunayana and Barai (2017) observed a water absorption of up to 5% for the coarse RCA obtained from a recycling plant, as mentioned in Table 2.1. They reported that smaller size particles with higher surface area and attached mortar exhibit higher absorption potential than larger size particles. The literature also documented similar relationship between particle size and water absorption, as mentioned in Table 2.2.

2.4.4 *Crushing Strength, Impact Strength, and Abrasion Resistance*

Mechanical properties like crushing, impact, and abrasion resistance are regulated by the porosity of aggregates. The crushing value of an aggregate is directly proportional to its strength, with a higher value indicating greater strength (Neville 1981). Studies have consistently shown that coarse RCA has a lower crushing value compared to coarse NCA. The 10% crushing value of coarse RCA was measured as 117 kN (Poon et al. 2009) and 120.5 kN (Rao et al. 2011). In comparison, the crushing value of NCA was 159 kN and 231.3 kN, respectively. The crushing method also has an influence on the crushing strength of coarse RCA. Primary crushing using jaw crusher produces RCA with lower crushing value as compared to the case where secondary crushing is used following the primary one (Li et al. 2009).

Impact test can be performed on aggregates in which a standardized hammer is dropped 15 times. It gives a measure of toughness of aggregates. A lower impact value of coarse aggregate corresponds to its higher toughness. Wong et al. (2007) reported that the impact values of RCA and NCA were 17.89 and 10.25, respectively. Rao et al. (2011) reported higher impact values of 35 in RCA and 17.37 in NCA. They suggested that, the reduced toughness of coarse RCA, as indicated by the higher impact value, may be the result of the crushing and removal of the adhered mortar layer during the impact test.

The Los Angeles abrasion test measures the percentage weight loss of aggregates, indicating their resistance to abrasion (Neville 1981). So, higher percentage loss refers to lesser abrasion resistance, a characteristic observed in coarse RCA. The Los Angeles abrasion value of RCA was experimentally determined to be 33.08, 37.2, 37.96, 42.7, and 37.1, as opposed to the values of NCA reported as 21.7, 24, 28.52,

Table 2.2 Density of coarse RCA and NCA

Author (year)	RCA source	Size (mm)	Bulk density (kg/m^3)		Density (SSD) (kg/m^3)		OD (kg/m^3)		WA (%)
			RCA	NCA	RCA	NCA	RCA	NCA	RCA
Eguchi et al. (2007)	Laboratory prepared CDW	20*	1260–1310						5.69–7.92
Poon et al. (2007)	Demolished concrete from site	10*					2490	2620	4.3
		20*					2570	2620	3.5
Tam et al. (2007)	Recycling plant	20							1.65
		10							2.63
Tangchirapat et al. (2008)	Laboratory prepared CDW	4.5–25			2450				5.61
González-Fonteboa and Martinez-Abella (2008)	Demolished concrete from site	4–12			2470	2650	2350	2640	4.82
		10–25			2480	2640	2370	2640	4.59
Gomes and de Brito (2009)	Laboratory prepared CDW(C30/37)	<25.4			2526	2573	2448	2550	8.49
Ferreira et al. (2011)	Demolish concrete waste	4–25.4	1140	1330	2440	2640	2300	2600	5.8
Fonseca et al. (2011)	Laboratory prepared CDW	4–25.4	1170	1440	2450	2580	2310	2550	6.1

* Maximum size

Table 2.3 Mechanical properties of coarse aggregate

Property	NCA	RCA	BIS limits
Los Angeles abrasion value (%)	14.4	34.08	50
Impact value (%)	13.87	24.23	45
Crushing value (%)	17.66	23.32	45

29.5, 21.56, respectively, in the following studies (Tangchirapat et al. 2008; López-Gayarre et al. 2009; Gomes and De Brito 2009; Poon et al. 2009; Rao et al. 2011). The abrasion resistance of coarse RCA may be enhanced by employing secondary crushing procedures with impact crushers (Li et al. 2009).

The mechanical properties of coarse aggregates, as shown in Table 2.3, were determined following the IS: 2386 (part-4) method by Sunayana and Barai (2017). The permissible values are also provided in accordance with this Indian standard, which indicates that the values are well within the limit for coarse RCA, even though they performed inferiorly compared to coarse NCA. The RCA was found to have inferior properties to the NCA in terms of impact resistance and crushing strength for two reasons: (1) rough surface texture of coarse aggregates and (2) presence of microcracks on the aggregate surface due to environmental exposure of parent concrete and crushing process. Higher abrasion value of RCA indicates lower resistance to abrasion. It is also attributed to the presence of weak residual mortar on the aggregate surface.

2.4.5 *Elastic Modulus*

One of the primary factors influencing the elastic modulus of concrete is the modulus of coarse aggregate (Duan et al. 2013). Specifically, the presence of the porous attached mortar on RCA surface has detrimental effects on the density and elastic modulus of aggregates as well as on their water absorption and crushing values (Liang et al. 2022). As the replacement level of coarse RCA increases, the modulus of elasticity of concrete decreases, assuming all other mix design parameters remain the same (Silva et al. 2014). This decrease is directly associated with the reduced elastic modulus of coarse RCA, which in turn affects the overall elasticity of the resulting concrete. According to Liang et al. (2022), the elastic modulus of RCA can be improved by strengthening the adhered mortar layer with fly ash-silica fume slurry or nanosilica. The pozzolanic materials not only act as filler to fill the microcracks in the old mortar layer but also react with unhydrated cement to produce more C-S-H gels.

2.4.6 *Alkali-Silica Reaction (ASR)*

The parameters such as source of parent concrete, presence of the residual mortar layer, recycling and crushing methods of coarse RCA cause variability in the properties of concrete. This also results in some complexity with respect to the susceptibility of coarse RCA to alkali-aggregate reaction. The factors responsible for ASR in coarse RCA-based concrete include residual mortar, size and density of aggregate, crushing stage, percentage replacement of RCA, alkali content, and saturation of RCA (Peng et al. 2020). The usage of RCA derived from ASR-affected concrete results in substantial expansion when incorporated into fresh concrete (Shehata et al. 2010; McCarthy et al. 2015). This expansion was comparable to that of the reactive aggregate that was originally employed in the old concrete. This higher reaction was attributed to the fact that, breaking up the old concrete revealed new surfaces for the reactive aggregate. This in turn stimulates new growth and expansion potential in the freshly mixed concrete.

The residual mortar may also release some alkalis into the pore solution causing more of this expansive reaction. The alkali content of cement as well as that of adhered mortar both can influence the rate of ASR. This action of alkali can be prevented by using washed coarse RCA in making concrete (Johnson and Shehata 2016). Residual mortar layer of coarse RCA may be responsible for promoting higher alkali-silica reaction due to its higher alkali reactivity (Etxeberria and Vázquez 2010). Nevertheless, the presence of reactive particles in the adhered mortar layer of RCA may be beneficial in limiting the exposure of parent aggregate, thereby reducing its susceptibility to this reaction. Few studies reported a different response of coarse RCA to the alkali reaction. The alkali-aggregate reaction may be lower in coarse RCA-based concrete, provided the reactive component in the parent aggregate has been already used up during the service life of the original concrete (Johnson and Shehata 2016). They also reported that smaller size coarse RCA (4.75–9.5 mm) are more prone to the ASR reaction due to higher reactive surface area leading to faster reaction unlike the larger size aggregates (9.5–12.5 mm). The coarse RCA produced from secondary crushing stage is more prone to ASR due to higher exposure of the reactive part of aggregates as compared to that in the aggregates obtained using primary crushing (Shehata et al. 2010; Adams et al. 2013). Increase of coarse RCA replacement from 25 to 100% results in an increase in ASR; it is also affected by the reactivity of parent aggregate (Beauchemin et al. 2018). Delobel et al. (2016) revealed that the ASR expansion of concrete produced from dried coarse RCA was reportedly more than that of concrete with water-saturated RCA. This is primarily due to the fact that dry RCA could cause more water absorption, thereby leading to the entry of more pore solution into the concrete.

2.4.7 *Thermal Properties of Aggregate*

The thermal expansion of concrete, which contains varying amounts of coarse RCA, increases as the temperature rises up to 800 °C (Zhao et al. 2018). This increase is mostly due to the thermal expansion of both the aggregate and the cement paste. They also observed that within the temperature range of 500 °C, the replacement % of coarse RCA has a negligible impact on the thermal expansion of concrete. The 50% replacement specimen exhibited the highest thermal expansion value beyond 500 °C, in contrast to the 0 and 100% coarse RCA concrete. The non-uniform thermal stresses between coarse NA and RCA may have contributed to the microcracking leading to higher thermal expansion in 50% RCA replaced concrete. At room temperature or beyond, the thermal conductivity of coarse RCA concrete decreases as the percentage replacement increases. At higher temperatures, the specific heat of concrete remains unaffected by the coarse RCA content.

2.5 Properties of Supplementary Cementitious Materials (SCMs)

The need for reducing cement consumption in concrete for environmental concerns has exploited the potential use of other cementitious materials as cement replacement. Ordinary Portland cement clinker with SCMs can reduce CO_2 emissions by 30–40% without substantially compromising the strength and durability properties (Skibsted and Snellings 2019). Most commonly used SCMs in concrete include fly ash (FA), ground granulated blast furnace slag (GGBFS), and silica fume (SF). The properties of these SCMs are discussed below.

2.5.1 *Fly Ash*

Fly ash (FA) is by-product obtained from burning of coal in coal-based electrical power plants. The physical and chemical properties of FA are regulated by the composition of raw coal, degree of pulverization, combustion temperatures, methods of collection, and storage conditions (Panesar and Zhang 2020; Ndahirwa et al. 2022) (Table 2.4).

Sunayana and Barai (2017) investigated the effect of coarse RCA and fly ash on the concrete properties for which they performed the chemical analysis of cement and fly ash following Indian Standard (IS: 4032-1985) and the results obtained are presented in Table 2.5. Combined percentage of SiO_2, Al_2O_3, and Fe_2O_3 in fly ash was amount to more than 70% of the total fly ash. It confirmed to the general requirements of Class F fly ash as per ASTM C-618.

Table 2.4 Physical properties of cement and fly ash (Sunayana and Barai 2017)

Property		Values
Consistency (%) (cement)		29
Fineness modulus (cement)		2.65
Specific gravity (cement)		3.13
Specific gravity (fly ash)		2.29
Specific surface area of cement (cm^2/g)		3060
Specific surface area of fly ash (cm^2/g)		3224.3
Setting time of cement (min)	Initial	75
	Final	295
Compressive strength of cement (MPa)	3 days	30.46
	7 days	44.4
	28 days	55.06

Table 2.5 Chemical composition of cement and fly ash

Materials	Compounds Compositions									
	L.O.I	SiO_2	Al_2O_3	Fe_2O_3	CaO	MgO	K_2O	Na_2O	TiO_2	SO_3
Cement	3.85	22.48	7.12	3.01	59.03	1.77	1.33	0.36	0.37	4.20
Fly ash	2.5	55.81	35.17	4.61	1.55	0.57	1.82	0.09	1.91	0.25

Table 2.6 Physical properties of SCMs (Ndahirwa et al. 2022)

SCMs	Particle size (μm)	Specific gravity	Specific surface area (m^2/kg)
Fly ash	5.89–23.4	1.3–2.9	250–600
Slag	10.8–14.4	2.6–2.9	340–600
Silica fume	0.1–58.2	2.1–2.5	150–30,000

2.5.2 *Slag*

Ground granulated blast furnace slag (GGBS) is a by-product from the iron manufacturing, obtained at temperatures about 1170–1500 °C. The physical properties are shown in Table 2.6 (Ndahirwa et al. 2022).

2.5.3 *Silica Fume*

Silica fume (SF) is produced as a by-product of the production of silicon and ferrosilicon alloys at a temperature of around 800 °C. It is ultrafine and mainly composed of amorphous silicon oxide which can have pozzolanic properties. Due to its finer particle size, it can contribute largely towards the improvement in filler effect, microstructural modification, pozzolanic reactivity, and consequently durability of concrete (Zhang et al. 2016).

2.6 Summary

The recycling methods for construction demolition wastes to produce coarse concrete aggregates have undergone significant development in due course of time. Simple methods, like mechanical crushing, non-explosive techniques, pressure jetting, thermal treatment, etc., to more sophisticated smart crushers, advanced dry recovery, thermochemical, thermomechanical methods are being used recently.

As the adhered mortar layer on the surface of coarse RCA is one of the prominent factors affecting the concrete properties, different surface treatment methods for reducing or strengthening this extra mortar layer are discussed. For the removal of attached mortar, mechanical or thermal methods are followed. For densifying this mortar layer, different surface coating with sodium silicates, bio-cement, silica fume slurry, accelerated carbonation, and bio-deposition methods may be applied.

Further, this chapter explored characterization of concrete constituents such as cementitious materials and aggregates. Though inferior to NCA, the physical and mechanical properties of RCA were well within the limit of usual ranges. Higher specific surface area of fly ash, silica fume, and slag implied greater fineness as compared to cement. After characterization of the materials, suitable mix proportioning method needs to be followed and applied to the SCM-based RAC, which are discussed in the next chapter.

References

Adams M, Jones A, Beauchemin S, Johnson R, Fournier B, Shehata M, Tanner J, Ideker J (2013) Applicability of the accelerated mortar bar test for alkali-silica reactivity of recycled concrete aggregates. Adv Civil Eng Mater 2(1):78–96

Al-Bayati HKA, Das PK, Tighe SL, Baaj H (2016) Evaluation of various treatment methods for enhancing the physical and morphological properties of coarse recycled concrete aggregate. Constr Build Mater 112:284–298

Alqarni AS, Abbas H, Al-Shwikh KM, Al-Salloum YA (2021) Treatment of recycled concrete aggregate to enhance concrete performance. Constr Build Mater 307:124960

Andreu G, Miren E (2014) Experimental analysis of properties of high performance recycled aggregate concrete. Constr Build Mater 52:227–235

Bairagi NK, Ravande K, Pareek VK (1993) Behaviour of concrete with different proportions of natural and recycled aggregates. Resour Conserv Recycl 9(1–2):109–126

Beauchemin S, Fournier B, Duchesne J (2018) Evaluation of the concrete prisms test method for assessing the potential alkali-aggregate reactivity of recycled concrete aggregates. Cem Concr Res 104:25–36

Berndt ML (2009) Properties of sustainable concrete containing fly ash, slag and recycled concrete aggregate. Constr Build Mater 23(7):2606–2613
Coelho A, de Brito J (2013) Environmental analysis of a construction and demolition waste recycling plant in Portugal-Part I: Energy consumption and CO_2 emissions. Waste Manage 33(5):1258–1267
de Andrade Salgado F, de Andrade Silva F (2022) Recycled aggregates from construction and demolition waste towards an application on structural concrete: a review. J Build Eng 52:104452
De Brito J, Ferreira J, Pacheco J, Soares D, Guerreiro M (2016) Structural, material, mechanical and durability properties and behaviour of recycled aggregates concrete. J Build Eng 6:1–16
Delobel F, Bulteel D, Mechling JM, Lecomte A, Cyr M, Rémond S (2016) Application of ASR tests to recycled concrete aggregates: influence of water absorption. Constr Build Mater 124:714–721
Dimitriou G, Savva P, Petrou MF (2018) Enhancing mechanical and durability properties of recycled aggregate concrete. Constr Build Mater 158:228–235
Duan ZH, Kou SC, Poon CS (2013) Using artificial neural networks for predicting the elastic modulus of recycled aggregate concrete. Constr Build Mater 44:524–532
Eguchi K, Teranishi K, Nakagome A, Kishimoto H, Shinozaki K, Narikawa M (2007) Application of recycled coarse aggregate by mixture to concrete construction. Constr Build Mater 21(7):1542–1551
Etxeberria M, Vázquez E (2010) Alkali silica reaction in concrete induced by mortar adhered to recycled aggregate. Mater Constr 60(297):47–58
Etxeberria M, Vázquez E, Marí A, Barra M (2007) Influence of amount of recycled coarse aggregates and production process on properties of recycled aggregate concrete. Cem Concr Res 37(5):735–742
Ferreira L, De Brito J, Barra M (2011) Influence of the pre-saturation of recycled coarse concrete aggregates on concrete properties. Mag Concr Res 63(8):617–627
Gebremariam AT, Di Maio F, Vahidi A, Rem P (2020) Innovative technologies for recycling End-of-Life concrete waste in the built environment. Resour Conserv Recycl 163:104911
Gokce A, Nagataki S, Saeki T, Hisada M (2011) Identification of frost-susceptible recycled concrete aggregates for durability of concrete. Constr Build Mater 25(5):2426–2431
Gomes M, de Brito J (2009) Structural concrete with incorporation of coarse recycled concrete and ceramic aggregates: durability performance. Mater Struct 42:663–675
Gómez-Soberón JM (2002) Porosity of recycled concrete with substitution of recycled concrete aggregate: an experimental study. Cem Concr Res 32(8):1301–1311
González-Fonteboa B, Martínez-Abella F (2008) Concretes with aggregates from demolition waste and silica fume. Materials and mechanical properties. Build Environ 43(4):429–437
Hansen TC (1985) Recycled aggregates and recycled aggregate concrete, 2nd state-of-the-art report developments 1945–1985. RILEM Technical Committee-37-DRC
Hilmersson S (1998) Hydrodemolition of concrete structures: basics and field experience. Water Jet Appl Constr Eng 163–176
Hossain MU, Poon CS, Lo IM, Cheng JC (2016) Comparative environmental evaluation of aggregate production from recycled waste materials and virgin sources by LCA. Resour Conserv Recycl 109:67–77
Ibrahim M, Alimi W, Assaggaf R, Salami BA, Oladapo EA (2023) An overview of factors influencing the properties of concrete incorporating construction and demolition wastes. Constr Build Mater 367:130307
IS (Indian Standard) (1963) Indian standards methods of tests for aggregates for concrete part I,III and IV, IS 2386–63, Bureau of Indian Standards, New Delhi, India (Reaffirmed 2002)
Johnson R, Shehata MH (2016) The efficacy of accelerated test methods to evaluate alkali silica reactivity of recycled concrete aggregates. Constr Build Mater 112:518–528
Katz A (2003) Properties of concrete made with recycled aggregate from partially hydrated old concrete. Cem Concr Res 33(5):703–711
Katz A (2004) Treatments for the improvement of recycled aggregate. J Mater Civ Eng 16(6):597–603

Kim HS, Kim B, Kim KS, Kim JM (2017) Quality improvement of recycled aggregates using the acid treatment method and the strength characteristics of the resulting mortar. J Mater Cycles Waste Manage 19:968–976

Kumbhar S, Gupta A, Desai D (2013) Recycling and reuse of construction and demolition waste for sustainable development. OIDA Int J Sustain Develop 6(7):83–92

Langer W (2016) Sustainability of aggregates in construction. In: Khatib JM (ed) Sustainability of construction materials. Woodhead publishing, Second Edition, pp 181–207

Li J, Xiao H, Zhou Y (2009) Influence of coating recycled aggregate surface with pozzolanic powder on properties of recycled aggregate concrete. Constr Build Mater 23(3):1287–1291

Li Y, Zhang S, Wang R, Zhao Y, Men C (2019) Effects of carbonation treatment on the crushing characteristics of recycled coarse aggregates. Constr Build Mater 201:408–420

Liang C, You J, Gu F, Gao Y, Yang G, He Z, Hou S, Duan Z (2022) Enhancing the elastic modulus of concrete prepared with recycled coarse aggregates of different quality by chemical modifications. Constr Build Mater 360:129590

López-Gayarre F, Serna P, Domingo-Cabo A, Serrano-López MA, López-Colina C (2009) Influence of recycled aggregate quality and proportioning criteria on recycled concrete properties. Waste Manage 29(12):3022–3028

McCarthy MJ, Csetenyi LJ, Halliday JE, Dhir RK (2015) Evaluating the effect of recycled aggregate on damaging AAR in concrete. Mag Concr Res 67(11):598–610

Mehta PK, Monteiro P (2006) Concrete: microstructure, properties, and materials

Mistri A, Dhami N, Bhattacharyya SK, Barai SV, Mukherjee A, Biswas WK (2021) Environmental implications of the use of bio-cement treated recycled aggregate in concrete. Resour Conserv Recycl 167:105436

Nagataki S, Gokce A, Saeki T, Hisada M (2004) Assessment of recycling process induced damage sensitivity of recycled concrete aggregates. Cem Concr Res 34(6):965–971

Ndahirwa D, Zmamou H, Lenormand H, Leblanc N (2022) The role of supplementary cementitious materials in hydration, durability and shrinkage of cement-based materials, their environmental and economic benefits: a review. Cleaner Materials 5:100123

Neville AM (1981) Properties of concrete, vol 4. Longman, London

Otsuki N, Miyazato SI, Yodsudjai W (2003) Influence of recycled aggregate on interfacial transition zone, strength, chloride penetration and carbonation of concrete. J Mater Civ Eng 15(5):443–451

Pandurangan K, Dayanithy A, Prakash SO (2016) Influence of treatment methods on the bond strength of recycled aggregate concrete. Constr Build Mater 120:212–221

Panesar DK, Zhang R (2020) Performance comparison of cement replacing materials in concrete: limestone fillers and supplementary cementing materials–a review. Constr Build Mater 251:118866

Pedro D, De Brito J, Evangelista L (2015) Performance of concrete made with aggregates recycled from precasting industry waste: influence of the crushing process. Mater Struct 48:3965–3978

Peng Z, Shi C, Shi Z, Lu B, Wan S, Zhang Z, Chang J, Zhang T (2020) Alkali-aggregate reaction in recycled aggregate concrete. J Clean Prod 255:120238

Poon CS, Shui ZH, Lam L, Fok H, Kou SC (2004) Influence of moisture states of natural and recycled aggregates on the slump and compressive strength of concrete. Cem Concr Res 34(1):31–36

Poon CS, Kou SC, Lam L (2007) Influence of recycled aggregate on slump and bleeding of fresh concrete. Mater Struct 40:981–988

Poon CS, Kou SC, Wan HW, Etxeberria M (2009) Properties of concrete blocks prepared with low grade recycled aggregates. Waste Manage 29(8):2369–2377

Raman JVM, Ramasamy V (2021) Augmentation of dissimilar techniques for enhancing the concrete properties with recycled coarse aggregate and manufactured sand. J Market Res 14:1180–1190

Rao MC, Bhattacharyya SK, Barai SV (2011) Behaviour of recycled aggregate concrete under drop weight impact load. Constr Build Mater 25(1):69–80

Rao MC, Bhattacharyya SK, Barai SV (2019) Systematic approach of characterization and behaviour of recycled aggregate concrete. Springer Trans Civil Environ Eng

Ravindrarajah R, Tam CT (1985) Properties of concrete made with crushed concrete as coarse aggregate. Mag Concr Res 37(130):29–38

Santha Kumar G, Minocha AK (2018) Studies on thermo-chemical treatment of recycled concrete fine aggregates for use in concrete. J Mater Cycles Waste Manage 20:469–480

Santha Kumar G, Saini PK, Karade SR, Minocha AK (2019) Chemico-thermal treatment for quality enhancement of recycled concrete fine aggregates. J Mater Cycles Waste Manage 21:1197–1210

Santos J, Branco FA, de Brito J (2002) Use of coarse recycled concrete aggregates in the production of new concrete. Structures 227–236

Sasanipour H, Aslani F (2020) Durability assessment of concrete containing surface pretreated coarse recycled concrete aggregates. Constr Build Mater 264:120203

Shehata MH, Christidis C, Mikhaiel W, Rogers C, Lachemi M (2010) Reactivity of reclaimed concrete aggregate produced from concrete affected by alkali–silica reaction. Cem Concr Res 40(4):575–582

Shima H, Tateyashiki H, Matsuhashi R, Yoshida Y (2005) An advanced concrete recycling technology and its applicability assessment through input-output analysis. J Adv Concr Technol 3(1):53–67

Silva RV, De Brito J, Dhir RK (2014) Properties and composition of recycled aggregates from construction and demolition waste suitable for concrete production. Constr Build Mater 65:201–217

Silva RV, De Brito J, Dhir RK (2017) Availability and processing of recycled aggregates within the construction and demolition supply chain: a review. J Clean Prod 143:598–614

Skibsted J, Snellings R (2019) Reactivity of supplementary cementitious materials (SCMs) in cement blends. Cem Concr Res 124:105799

Spaeth V, Tegguer AD (2013) Improvement of recycled concrete aggregate properties by polymer treatments. Int J Sustain Built Environ 2(2):143–152

Sunayana S, Barai SV (2017) Recycled aggregate concrete incorporating fly ash: comparative study on particle packing and conventional method. Constr Build Mater 156:376–386

Tam VW, Tam CM, Wang Y (2007) Optimization on proportion for recycled aggregate in concrete using two-stage mixing approach. Constr Build Mater 21(10):1928–1939

Tangchirapat W, Buranasing R, Jaturapitakkul C, Chindaprasirt P (2008) Influence of rice husk–bark ash on mechanical properties of concrete containing high amount of recycled aggregates. Constr Build Mater 22(8):1812–1819

Tangchirapat W, Buranasing R, Jaturapitakkul C (2010) Use of high fineness of fly ash to improve properties of recycled aggregate concrete. J Mater Civ Eng 22(6):565–571

Topçu IB, Guncan NF (1995) Using waste concrete as aggregate. Cem Concr Res 25(7):1385–1390

Vieira JPB, Correia JR, De Brito J (2011) Post-fire residual mechanical properties of concrete made with recycled concrete coarse aggregates. Cem Concr Res 41(5):533–541

Wang B, Yan L, Fu Q, Kasal B (2021) A comprehensive review on recycled aggregate and recycled aggregate concrete. Resour Conserv Recycl 171:105565

Wong YD, Sun DD, Lai D (2007) Value-added utilisation of recycled concrete in hot-mix asphalt. Waste Manage 27(2):294–301

Wu CR, Hong ZQ, Zhang JL, Kou SC (2020) Pore size distribution and ITZ performance of mortars prepared with different bio-deposition approaches for the treatment of recycled concrete aggregate. Cement Concr Compos 111:103631

Zhang J, Taylor PC, Shi C (2015) Investigation of approaches for improving interfacial transition zone-related freezing-and-thawing resistance in concrete pavements. ACI Mater J 112(5):613–617

Zhang Z, Zhang B, Yan P (2016) Comparative study of effect of raw and densified silica fume in the paste, mortar and concrete. Constr Build Mater 105:82–93

Zhang H, Ji T, Liu H, Su S (2021) Improving the sulfate resistance of recycled aggregate concrete (RAC) by using surface-treated aggregate with sulfoaluminate cement (SAC). Constr Build Mater 297:123535

Zhao H, Liu F, Yang H (2018) Thermal properties of coarse RCA concrete at elevated temperatures. Appl Therm Eng 140:180–189

Chapter 3
Mixing and Mix Proportioning Methods

Preview

Concrete performance depends not only on the constituent materials but also on the proportions in which the materials are mixed and the order of mixing during casting. So, the constituents of known qualities need to be proportioned accordingly to achieve concrete of desired performance. This chapter explores the utilization of the particle packing method for mix design, specifically for determining the appropriate proportions of fly ash and coarse recycled concrete aggregate (RCA) in concrete. The effects of mixing methods such as two-stage and triple mixing are also discussed in this context.

3.1 Material Mixing Methods and Their Significance

When mixing concrete constituents in the mixer, factors such as the sequence in which materials are added, the amount of time the mix is left to mix, the total energy required to mix, and the technique of discharging the mix influence the concrete behaviour (Ferraris 2001). In this context, the two-stage mixing approach and triple mixing methods are discussed for coarse RCA-based concrete in this section.

3.1.1 Conventional Method

The conventional mixing of ingredients during concrete production process involves dry mixing followed by wet mixing (Charonnat and Beitzel 1997). Dry mixing refers to the process of mixing of cement, fine aggregates, and coarse aggregates that takes place during loading, prior to the introduction of water. Wet mixing refers to the process of mixing that occurs either after or simultaneously with the addition of

S. Sunayana and S. V. Barai, *Green Concrete*, Springer Transactions in Civil and Environmental Engineering, https://doi.org/10.1007/978-981-96-6982-0_3

water, while the loading is still in progress. This indicates that materials can be introduced at any point throughout the loading period; before addition of water, after water, or a combination of both.

3.1.2 Two-Stage Mixing Method

Tam et al. (2007) devised the two-stage mixing approach (TSMA). In this method, initially half of the total estimated quantity of water is introduced to the dry mixed aggregates and then cement is added. As a result, a coating of cement slurry is formed on the surface of RCA. The slurry permeated the weak adhered mortar layer and filled pre-existing microcracks in RCA. This method leads to the formation of a greater quantity of specific chemical products, including calcium silicate hydrate, portlandite, and ettringite, which in turn contribute to the development of a more stronger interfacial transition zone (ITZ).

3.1.3 Triple Mixing Method

Kong et al. (2010) introduced triple mixing method for RAC, which resulted in additional improvements in the microstructure and strength. In this method, coarse aggregates and sand are added to the mixer with a set amount of water and mixed for 15 s. Afterwards, mineral admixtures such as fly ash is introduced to the moist aggregates and again blended for an additional 15 s to coat the aggregate surfaces. Then, cement is added to the coated aggregates and the mixing continues for 30 s. Finally, the balance quantity of water and superplasticizer are combined and stirred for 60 s. This approach shows enhancements in strength and other concrete properties.

Sunayana and Barai (2017) adopted triple mixing method for feeding the ingredients into the mixer for fly ash-based RAC as follows. The mixing method is illustrated in Fig. 3.1. Aggregates of different sizes, in the proportion obtained from the design mix, were mixed in dry state in the mixer. Then the dry mixed aggregates were subjected to a slurry made from a specific amount of water (W1) in fly ash. This method forms a coating of fly ash slurry on the surface of coarse recycled concrete aggregates.

After proper mixing of the coated aggregates, cement is introduced and blended first in dry condition for 30 s and then for 60 s with the addition of remaining water to generate the newly mixed concrete. Based on the trial, the ratio of W1 to fly ash is maintained at 0.69 and 0.46 for RAC with 20% and 30% fly ash replacement, respectively. Superplasticizer is added to the water at a dosage of 0.37% and 0.42% of total cementitious material respectively, to produce RFA20 and RFA30 concrete. The proportion of superplasticizer is determined through various trials to achieve the target workability of 100 $\pm$ 20 mm slump.

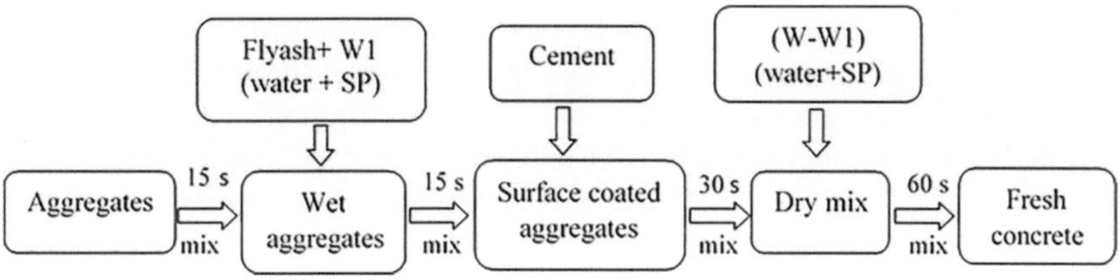

Fig. 3.1 Mixing method used for FARAC

3.2 Mix Design Methods

Proportioning of concrete constituents influences properties of structural concrete significantly. Thus, implementation of a proper mix design method is highly essential for any type of concrete. Different mix design methods are discussed below.

3.2.1 *Conventional Mix Design*

For a specified replacement ratio, the direct weight replacement method (Dhir et al. 1999; Gomez-Soberon 2002) presumes that the total weight of coarse aggregate (RCA + NCA), cement, and water would remain constant. For the same volume in both RAC and NAC, the weight of fine aggregate is only altered, specifically decreased, due to RCA. In the direct volume replacement method (Poon et al. 2004; Salem et al. 2003), the lower specific gravity of RCA is accounted for by replacing similar volume of coarse NCA by RCA to produce concrete. As long as the overall volume of coarse aggregates remains constant, the quantities of fine aggregate, cement, and water need not be changed. However, these methods were unable to counteract the detrimental impact of RCA on the properties of resulting concrete. As mentioned earlier, the inferiorities caused in RAC are mainly attributed to the old mortar content on RCA surface and its higher porosity. Other modifications such as adjustment of water-to-cement (w/c) ratio or water content and pre-soaked coarse aggregates were also adopted during mix designs as reported in few studies (Topcu and Sengel 2004; Sagoe-Crentsil et al. 2001; Dhir et al. 1999). Using a higher cement content to compensate for the attached mortar in RAC, as suggested by Etxeberria et al. (2007), proved to be an ineffective solution; However, there may be an increase in shrinkage, creep, and adverse environmental concerns related to cement.

3.2.2 *Equivalent Volume Method*

Fathifazal et al. (2009) suggested the equivalent mortar volume (EMV) method of mix design for RAC. It assumed that the combined amount of old and fresh mortar in

RAC needs to be identical as that of total mortar contents in NAC. The attributes of fresh concrete, such as density and workability, as well as the properties of hardened concrete, such as compressive strength and modulus of elasticity, were enhanced in the case of RAC as compared to the conventional mix design. However, this method required precise determination of adhered mortar content which adds complexities to its feasible application. Various methods such as nitric acid dissolution (Movassaghi 2006), image analysis (Abbas et al. 2008), and thermal treatment (de Juan and Gutiérrez 2004), chemical treatment or applying freeze–thaw cycles may be used for estimation of the residual mortar quantity. But, the necessity of such pre-treatment steps for RCA in the EMV approach posed challenges for its practical implementation. In addition, 100% RCA replacement was not possible, if the actual adhered mortar content was higher than the maximum permissible residual mortar based on specific gravity of RCA, NCA, and dry volume of NCA in NAC. The above-mentioned issues highlight the necessity for a novel mix design approach for 100% coarse RCA replaced concrete. The particle packing method (PPM) is intended to serve as a suitable mix design approach in RAC due to its underlying philosophies as discussed in the following sections.

3.2.3 *Particle Packing Method*

Packing density (PD) refers to the proportion of the volume occupied by solid material to the overall volume. The best possible combination of different-sized aggregates targeting maximum packing density with minimization of voids can be achieved by PPM. Methods for determination of packing density can be categorized into the following three types:

3.2.3.1 Using Optimization Curves

Idealized optimization curves such as Fuller curve (Fuller and Thompson 1907) or Andersen curve (Andreasen and Andresen 1930) can be used for particle packing.

Sunayana and Barai (2017) applied the modified Andersen curve for optimization of RAC mix. Initially, proportioning of aggregates that is finding out a proper gradation for concrete constituents is accomplished using the ideal optimization curve. These curves are based on the continuous particle size distribution considering geometry of the particles. For this purpose, modified Andersen curve is considered here (Eq. 3.1).

$$P(d) = \frac{d^q - {d_{min}}^q}{{d_{max}}^q - {d_{min}}^q} \tag{3.1}$$

$P(d)$ size cumulative distribution function.

d particle diameter under consideration.

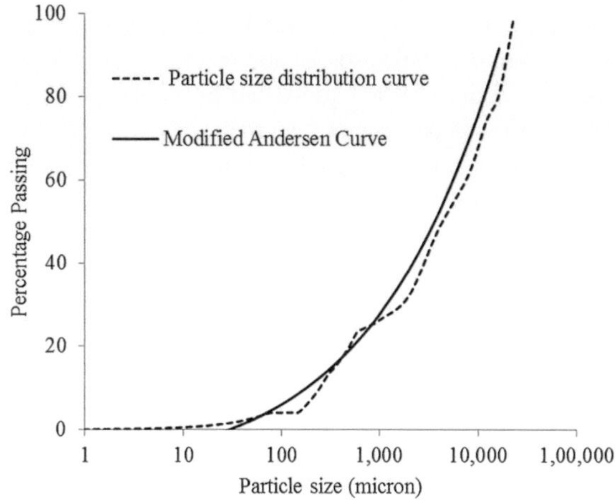

Fig. 3.2 Comparison between theoretical and modified Andersen curve

d_{max} maximum particle diameter in the mixture.
d_{min} minimum particle diameter in the mixture.
q distribution coefficient to adjust the curve for fineness.

The EMMA mix analyzer is employed to generate the particle size distribution curve of RAC. The coefficient q is set at 0.35 based on trials to align the derived curve with the Andersen curve (Fig. 3.2). A good match is observed for the assumed particle size proportions of RAC. Nevertheless, this technique solely considers the size of the particles, disregarding their shape, properties, and surface texture of aggregates (which is particularly important in the context of RCA). Therefore, this technique may or may not yield the highest possible PD.

3.2.3.2 Using Analytical Particle Packing Models

Analytical models such as modified Toufar model (Goltermann et al. 1997), Dewar model (Dewar 1999), linear packing (Stovall et al. 1986), and compressible packing model (CPM) (de Larrard 1999) have been used previously for particle packing. These models are used for optimization of larger size aggregates (>125 μm). CPM is derived from the linear packing model and considers wall effect and loosening effects in packing. Compaction index is used to calculate the actual PD from the virtual PD. Parameters such as the pouring method, particle shape, density, and compaction energy have an impact on the packing of aggregates.

These theoretical packing approaches are found to be unsuitable for packing of finer materials, such as cement or fillers with larger surface areas, even if they could account for the geometry of larger size particles. In a study conducted by Jian (2012), it was found that finer materials, which have a larger surface area to volume ratio, effectively occupy the empty spaces between cement particles and enhance the density of the paste. In contrast to coarse aggregates, where simple geometrical interactions are predominant, fine particles also exhibit strong surface-related van der Waals forces and electrostatic charges. The compaction-interaction packing model (CIPM) suggested by Fennis et al. (2013) considers these interactions aspects in packing of finer materials. Computation of packing density considers the effect of both aggregates and fillers. The interaction coefficients used in CPM model are modified to account for the surface-related forces arising in finer particles. It was thought that these forces are responsible for the inter-particle friction. The compaction energy used in CPM is modified due to the reduced tendency of agglomeration and increased compaction of fine particles. The mathematical equations are discussed in the following section. A higher packing density indicates fewer empty spaces or voids that need to be filled with water.

Analytical models that account for the interaction between particles of varying geometries was implemented by Sunayana and Barai (2017) to ascertain the optimum PD. The compaction-interaction packing model was employed to determine the virtual packing density of a mix with predominant size class 'i' (Eq. 3.2).

$$\beta_{ti} = \frac{\beta_i}{1 - \sum_{j=1}^{i-1}\left[1 - \beta_i + b_{ij}\beta_i\left(1 - 1/\beta_j\right)\right]r_j - \sum_{j=i+1}^{n}\left[1 - a_{ij}\beta_i/\beta_j\right]r_j} \tag{3.2}$$

The packing density of mono-sized particles (α_j) was used to calculate the virtual packing density of a specific size class (β_j) as given in Eq. 3.3. k denotes compaction energy.

$$\beta_j = \left(1 + \frac{1}{k}\right) \times \alpha_j. \tag{3.3}$$

When the larger sized particles are predominant in a mixture, and smaller particles are not sufficiently small to accommodate in between large particles, then the packing gets affected by a phenomenon known as 'loosening effect'. Similarly, when fine particles are predominant in a mix, and if larger particles are not large enough to fit in their gap, disrupt the packing by generating extra voids near the wall of coarse particles. This is recognized as the 'wall effect'. Equations 3.4 and 3.5 in the CPM model provide the interaction coefficients for loosening and wall effects. Compaction energy can be applied to increase effectiveness of packing (Eq. 3.6). The CPM model only considers interaction of larger size (>125 μm) particles in packing.

Coefficient representing loosening effect

$$a_{ij} = \sqrt{1 - (1 - d_j/d_i)^{1.02}}. \tag{3.4}$$

Coefficient representing wall effect

$$b_{ij} = 1 - (1 - d_j/d_i)^{1.5}. \tag{3.5}$$

Compaction index

$$k = \sum k_i = \sum_{i=1}^{n} \frac{r_i/\beta_i}{(1/\alpha_t - 1/\beta_{ti})}. \tag{3.6}$$

To account for the packing of smaller particles such as cement along with other cementitious material used as cement replacement, where surface forces play the predominant role, CIPM can be applied. Interaction coefficients (a_{ij} and b_{ij}) in CIPM are expressed below.

$$a_{ij} = \left\{ \begin{matrix} 1 - \left[\frac{\log_{10}\frac{d_i}{d_j}}{w_{0,a}}\right] if \log_{10}\frac{d_i}{d_j} < w_{0,a} \\ 0 if \log_{10}\frac{d_i}{d_j} > w_{0,a} \end{matrix} \right\}. \tag{3.7}$$

Function for maximum range of loosening effect

$$w_{0,a} = \left\{ \begin{matrix} w_a \times c_a if d_j < d_c \\ w_a if d_j \geq d_c \end{matrix} \right\} \tag{3.8}$$

$$b_{ij} = \left\{ \begin{matrix} 1 - \left[\frac{\log_{10}\frac{d_j}{d_i}}{w_{0,b}}\right] if \log_{10}\frac{d_j}{d_i} < w_{0,b} \\ 0 if \log_{10}\frac{d_j}{d_i} > w_{0,b} \end{matrix} \right\} \tag{3.9}$$

$$w_{0,b} = \left\{ \begin{matrix} w_b \times c_b if d_i < d_c \\ w_b if d_i \geq d_c \end{matrix} \right\}. \tag{3.10}$$

The parameters used for cement with fly ash substitution in the presence of superplasticizer may be considered as $w_a = w_b = 1$, $c_a = 1.5$, $c_b = 0.2$ and $d_c = 25$ μm (Fennis et al. 2013). The MATLAB program was used to calculate theoretical PD using CPM and CIPM. The maximum PD obtained using the CIPM method was 0.85 which is significantly higher (16%) than the experimental PD computed in the following section. By utilizing the CPM model, which does not consider the interaction effect between extremely small particles, a maximum packing density of 0.74 was obtained. This condition resembles the experimental situation for measuring packing density; hence, the theoretical PD comes closer to the experimental value. The experimental PD calculation is detailed in the following section. Theoretical particle density (PD) refers to the hypothetical arrangement of aggregates stacked on top of each other, which always results in a closer packing and higher packing density than the actual experimental PD.

3.2.3.3 Using Experimental Methods

Jacobsen and Arntsen (2008) performed experiments for varying packing density of aggregates. Particle size distribution, shape of aggregates, and interaction effects can influence the PD. Kwan et al. (2013) reported the role of moisture content of aggregate on the PD. Effective blending of fine and coarse aggregates may be facilitated under wet condition of the particles resulting in higher PD. Raj et al. (2014) employed PPM mix design for different grades of NAC. They observed comparable workability and 28-day compressive strength for PPM as well as the Indian standard (IS) method of mix design.

Sunayana and Barai (2017) applied this PPM method experimentally to fly ash incorporated 100% coarse RCA-based concrete as discussed below. PPM is method of arranging aggregates where smaller aggregates are placed within the gaps between larger sizes to minimize the voids and maximize the density of packing. The presence of excess mortar on the RCA surface is automatically taken into consideration while determining the remaining voids that can be filled with fresh mortar. As a result, the requirement of fresh paste content is reduced.

Experimental trials are conducted to determine packing density using various coarse aggregate sizes and proportions. The bulk density of coarse aggregates with sizes of 20 and 12.5 mm was measured for various mass proportions, such as 90:10, 80:20, 70:30, and so forth. The packing density of different trial mixture combinations was determined using the bulk density using Eq. 3.11. The proportion of different-sized aggregates that achieved the highest packing density is obtained. To this proportion of 12.5 to 20 mm sized aggregates, 10 mm size is added in different trial mixes (90:10, 80:20, 70:30, and so on). The aforementioned procedures are then repeated for 6.3 mm aggregates in order to determine the proportions of all the involved aggregates that yield the highest possible PD. Then with this finalized coarse aggregate proportion, fine aggregates are added. A similar exercise is used to determine the coarse-to-fine aggregate ratio for the maximum possible bulk density and the corresponding maximum PD as per Eq. 3.11.

$$\textit{Packing density} = \sum \frac{\textit{bulk density} \times \textit{weight fraction}}{\textit{specific gravity}}. \tag{3.11}$$

The results of different trials for obtaining the maximum packing density are displayed in Fig. 3.3. It is found that the bulk density and PD grow initially, reach their maximum, and then start to decline. This behaviour occurs because the presence of an excessive amount of smaller particles (beyond the maximum limit) disrupts the arrangement of larger particles, resulting in a decrease in the packing density. In this manner, the optimal weight proportions of various aggregate sizes that correspond to the maximum PD are determined, as well as the maximum packing density. The PD of the mixture is calculated solely using the coarse aggregates and fine aggregate. The incorporation of finer proportions, such as smaller-sized coarse aggregates and fine aggregates, fills the spaces in between larger coarse aggregates, resulting in a higher packing density of the mix. The calculations are shown as follows.

In the present experimental trials, maximum bulk density $= 1.865$ g/cm^3.

Maximum packing density $= 1.865 \times \left[\frac{0.3192}{2.54} + \frac{0.1368}{2.53} + \frac{0.114}{2.43} + \frac{0.03}{2.29} + \frac{0.4}{2.66}\right] = 0.728$.

This PD value obtained was around 2.4% lower than the theoretical PD (0.74). Furthermore, to enhance packing of the paste, cement can be substituted partly with

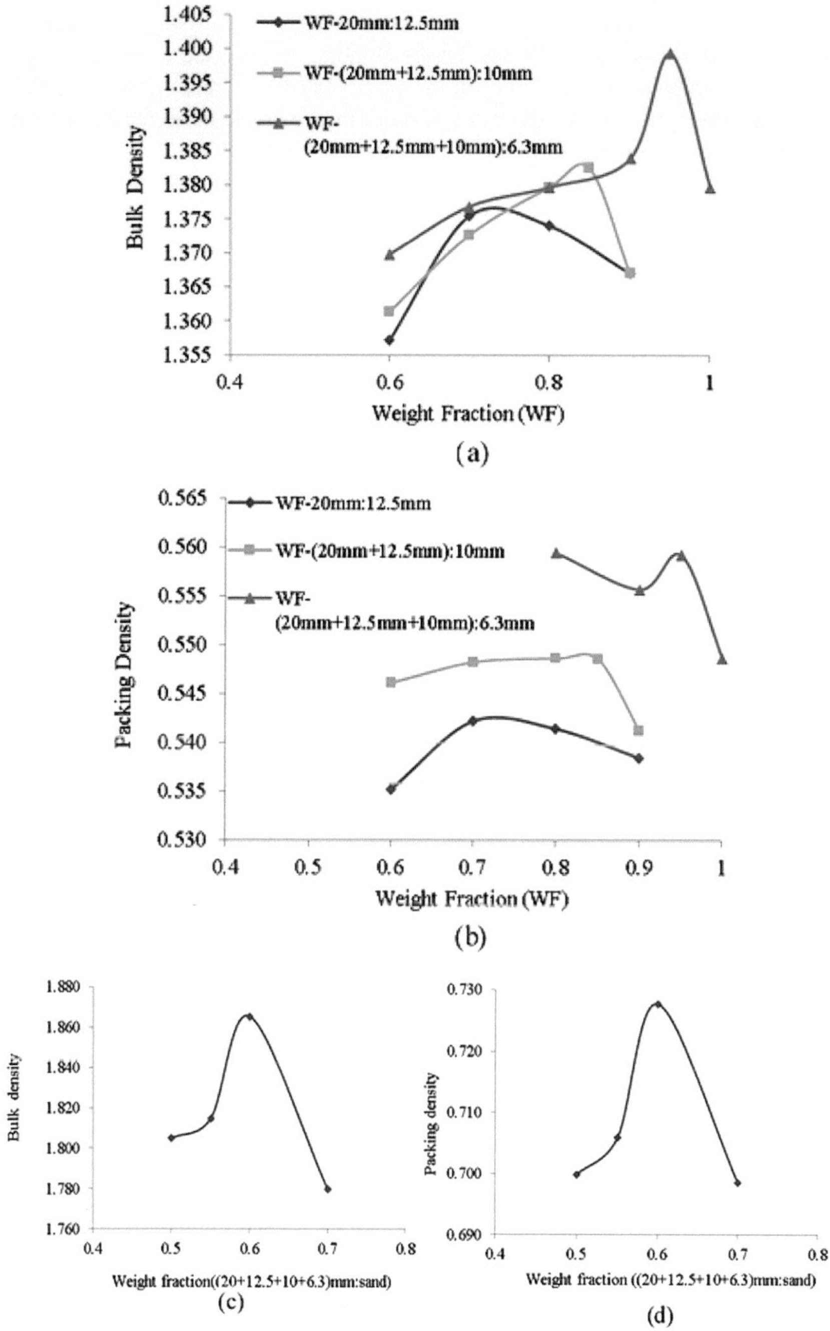

Fig. 3.3 Calculation of maximum packing density: **a** experimental bulk density vs. weight fraction for different trials, **b** packing density vs. weight fraction corresponding to bulk density, **c** final trial proportions for bulk density, **d** final weight fraction for maximum packing density

any other cementitious material finer than it that is particles with larger surface area (Jian 2012). In this study performed by the authors, fly ash used as partial cement substitution further contributed towards increase of PD of concrete. Although in PPM mix design, the decrease in voids content resulted in an increase in aggregate amounts, the reduction in binder requirement was not significant. Instead, the quantity of binder materials was selected based on their capacity to provide acceptable workability and prevent the formation of harsh concrete.

The calculating stages were carried out in the following manner:

Void content $= 1 - 0.728 = 0.272$.

Thus, paste content was calculated as 16% in the excess of voids based on trials.

Paste content $= 0.272 + 0.272 \times 0.16 = 0.3155$.

For M30 grade of concrete, w/c ratio was fixed based on trial mixes to satisfy the workability and strength criteria.

$w/c = 0.45, w = 0.45c$

Total paste $= C + w = \frac{c}{3.13} + \frac{0.45c}{1} = 0.769c$

Cement content $= \frac{0.3155}{0.769} \times 1000 = 410.27\text{kg/m}^3$

Cementitious material content (with 10% increase based on trials) $= 410.27 \times 1.1 = 451.29\ \text{kg/m}^3$

For fly ash replacement of 20%,

Total paste $= c + w = \frac{0.8c}{3.13} + \frac{0.2c}{2.29} + \frac{0.45c}{1} = 0.793\text{c}$

Paste content $= 451.29 \times \frac{0.793}{1000} = 0.358$

Volume of aggregate $= 1 - 0.358 = 0.6424$

Total solid volume of aggregate $= \sum \frac{Weightfractionofaggregate}{specificgravity}$

$= \frac{0.3192}{2.54} + \frac{0.1368}{2.53} + \frac{0.114}{2.43} + \frac{0.03}{2.29} + \frac{0.4}{2.66}$

$= 0.39$

Weight of 20 mm aggregates $= \frac{0.6424}{0.39} \times 0.3192 \times 1000 = 525.78\text{kg/m}^3$

Weight of 12.5 mm aggregates $= \frac{0.6424}{0.39} \times 0.1368 \times 1000 = 225.33\text{kg/m}^3$

Weight of 10 mm aggregates $= \frac{0.6424}{0.39} \times 0.114 \times 1000 = 187.78\text{kg/m}^3$

Weight of 6.3 mm aggregates $= \frac{0.6424}{0.39} \times 0.03 \times 1000 = 49.42\text{kg/m}^3$

Weight of sand $= \frac{0.6424}{0.39} \times 0.4 \times 1000 = 658.87\ \text{kg/m}^3$

Water content $= 0.45 \times 451.04 \times 0.95 = 192.82\text{kg/m}^3$

Further adjustments were made to account for water absorption and moisture content to determine the precise quantity of water and aggregates needed. Weight of aggregate proportions obtained by optimizing the packing density of the mix is shown in Table 3.1. The control concrete, designated as NAC, consisted of natural coarse and fine aggregates and cement as the binder ingredient, without any fly ash replacement. RFA20 refers to a type of concrete that is made up of 100% coarse recycled concrete aggregates, natural fine aggregates, and has 20% of its cement replaced with fly ash. Similarly, RFA30 concrete comprised of 100% coarse RCA, natural fine aggregates, and cement with 30% fly ash replacement. The final proportion of concrete constituents calculated following PPM and conventional Indian standard method are shown in Table 3.2.

The reasons for variation in quantities of the RCA and NCA are the difference in their specific gravity. A comparison between PPM and conventional methods (IS

Table 3.1 Weight of aggregates obtained using PPM (Sunayana and Barai 2017)

Aggregate size	20–12.5 mm	12.5–10 mm	10–6.3 mm	6.3–4.75 mm	Sand
Weight (kg/m^3)	525.78	225.33	187.78	49.42	658.87

Table 3.2 Mix design proportions (Sunayana and Barai 2017)

Mix type	Method	w/b	Cement kg/m^3	Fly ash kg/m^3	NCA		Sand kg/m^3	RCA	
					(20–10) mm kg/m^3	(10–6.3) mm kg/m^3		20–10 mm kg/m^3	10–6.3 mm kg/m^3
RFA20	PPM	0.45	361	90	–	–	659	764	243
RFA30	PPM	0.45	316	135	–	–	659	762	243
NAC	PPM	0.45	410	–	914	229	762		
NAC	IS code	0.45	438	–	756	486	610	–	–
RFA20	IS code	0.45	372	93	–	–	600	664	415
RFA30	IS code	0.45	325	139	–	–	600	664	415

provisions) at a constant water-to-binder ratio revealed significant difference in the quantity of fine aggregates. The PPM approach utilizes fine particles to enhance the PD, thereby yielding a mixture with a relatively greater proportion of sand. Due to increase in total aggregate proportions, the voids that need to be filled with paste are reduced resulting in lowering the requirement of cementitious materials.

3.3 Summary

This chapter discussed concrete mixture proportioning methods with a detail focus on the particle packing method (PPM) for concrete (designated as FARAC) comprising fly ash and coarse recycled concrete aggregate. Initially, the packing density of aggregates was estimated by both theoretical and experimental methods. The weight fraction of aggregates of various sizes that corresponded to the maximum packing density was utilized in the preparation of FARAC. The cement content in FARAC was minimized due to the combined influence of PPM and fly ash. Triple mixing method was applied for mixing the constituents for concrete production. The subsequent chapters will explore the impact of PPM and triple mixing on FARAC, focusing on micro, macro, and structural components.

References

Abbas A, Fathifazl G, Isgor OB, Razaqpur A, Fournier B, Foo S (2008) Proposed method for determining the residual mortar content of recycled concrete aggregates. J ASTM Int 5(1):1–12

Andreasen AHM, Andersen J (1930) About the relationship between grain gradation and gap in products of loose grains (with some experiments). Colloid Polym Sci 50(3):217–228

Charonnat Y, Beitzel H (1997) RILEM TC 150 ECM: Efficiency of concrete mixers; report: efficiency of concrete mixers to-wards qualification of mixers. Mater Struct (Suppl. 196) 30:28–32

De Larrard F (1999) Concrete mixture proportioning: a scientific approach. CRC Press

Dewar J (1999) Computer modelling of concrete mixtures. CRC Press

Dhir RK, Limbachiya MC, Leelawat T, BS 5328, BS 882 (1999) Suitability of recycled concrete aggregate for use in BS 5328 designated mixes. Proc Inst Civ Eng-Struct Build 134(3):257–274

Etxeberria M, Vázquez E, Marí A, Barra M (2007) Influence of amount of recycled coarse aggregates and production process on properties of recycled aggregate concrete. Cem Concr Res 37(5):735–742

Fathifazl G, Abbas A, Razaqpur AG, Isgor OB, Fournier B, Foo S (2009) New mixture proportioning method for concrete made with coarse recycled concrete aggregate. J Mater Civ Eng 21(10):601–611

Fennis SAAM, Walraven JC, Den Uijl JA (2013) Compaction-interaction packing model: regarding the effect of fillers in concrete mixture design. Mater Struct 46:463–478

Ferraris CF (2001) Concrete mixing methods and concrete mixers: state of the art. J Res Nat Inst Stand Technol 106(2):391

Fuller WB, Thompson SE (1907) The laws of proportioning concrete. Trans Am Soc Civ Eng 59(2):67–143

Goltermann P, Johansen V, Palbøl L (1997) Packing of aggregates: an alternative tool to determine the optimal aggregate mix. Mater J 94(5):435–443

Gómez-Soberón JM (2002) Porosity of recycled concrete with substitution of recycled concrete aggregate: an experimental study. Cem Concr Res 32(8):1301–1311

Jacobsen S, Arntsen B (2008) Aggregate packing and-void saturation in mortar and concrete proportioning. Mater Struct 41(4):703–716

Jian CJ (2012) Use of fillers to improve packing density and performance of concrete. Ph.D. thesis, The University of Hong Kong, China

De Juan MS, Gutierrez PA (2004) Influence of recycled aggregate quality on concrete properties. In: Conference on the use of the recycled materials in buildings and structures, Barcelona, Spain

Kong D, Lei T, Zheng J, Ma C, Jiang J, Jiang J (2010) Effect and mechanism of surface-coating pozzalanics materials around aggregate on properties and ITZ microstructure of recycled aggregate concrete. Constr Build Mater 24(5):701–708

Kwan AKH, Chan KW, Wong V (2013) A 3-parameter particle packing model incorporating the wedging effect. Powder Technol 237:172–179

Movassaghi R (2006) Durability of reinforced concrete incorporating recycled concrete as aggregate (RCA) (Master's thesis, University of Waterloo)

Poon CS, Shui ZH, Lam L (2004) Effect of microstructure of ITZ on compressive strength of concrete prepared with recycled aggregates. Constr Build Mater 18(6):461–468

Raj N, Patil SG, Bhattacharjee B (2014) Concrete mix design by packing density method. IOSR J Mech Civ Eng 11(2):34–46

Sagoe-Crentsil KK, Brown T, Taylor AH (2001) Performance of concrete made with commercially produced coarse recycled concrete aggregate. Cem Concr Res 31(5):707–712

Salem RM, Burdette EG, Jackson NM (2003) Resistance to freezing and thawing of recycled aggregate concrete. Mater J 100(3):216–221

Stovall T, De Larrard F, Buil M (1986) Linear packing density model of grain mixtures. Powder Technol 48(1):1–12

Sunayana S, Barai SV (2017) Recycled aggregate concrete incorporating fly ash: Comparative study on particle packing and conventional method. Constr Build Mater 156:376–386
Tam VW, Tam CM, Wang Y (2007) Optimization on proportion for recycled aggregate in concrete using two-stage mixing approach. Constr Build Mater 21(10):1928–1939
Topcu IB, Şengel S (2004) Properties of concretes produced with waste concrete aggregate. Cem Concr Res 34(8):1307–1312

Chapter 4
Environmental Impact of Concrete

Preview

Sustainable concrete production demands satisfaction of the two most important criteria such as optimization of resources and minimization of end-of-life wastes. As discussed in the previous chapters, one of the important ways in which this can be achieved is by using coarse recycled concrete aggregates (RCAs) as an alternative to natural coarse aggregates and other industrial wastes as supplementary cementitious materials (SCMs). In this regard, the role of a suitable mix design method for obtaining an optimum combination of concrete constituents is also significant. In this chapter, sustainability of 100% recycled coarse aggregate concrete (RAC) along with SCMs in concrete is explored through performing life-cycle assessment, especially focusing on concrete with coarse RCA and low-volume fly ash as cement substitute (denoted as FARAC) in concrete. The effect of particle packing method (PPM) of mix proportioning on the environmental impact of FARAC is assessed for a case study of India.

4.1 Environmental Impact of RAC

Environmental impact analysis of any product or process is basically carried out through life-cycle assessment following the standard ISO 14040. The systematic methodological approach of LCA involves four basic components as follows: (1) definition of goal and scope: identify functional unit and specify system boundary, (2) preparation of life-cycle inventory (LCI), (3) life-cycle impact assessment, and (4) interpretation of results.

The application of recycled coarse aggregates instead of natural aggregates in concrete is beneficial in terms of resource optimization. However, quantitative assessment of the net environmental impact, considering the recycling activities and transportation of raw materials, is an ongoing matter of investigation. For both RAC

S. Sunayana and S. V. Barai, *Green Concrete*, Springer Transactions in Civil and Environmental Engineering, https://doi.org/10.1007/978-981-96-6982-0_4

and NAC, the overall environmental effects, including energy consumption, contribution to global warming, eutrophication, acidification, and formation of photochemical oxidants, are influenced by both the distance and the mode of transportation (Marinković et al. 2010). For equal transport distance of RAC and NAC, they reported a higher adverse environmental impact in the later, which is in a range of 11–37% based on different impact categories. Cement is found to be the main factor responsible for environmental concerns of concrete.

The environmental impact of 216 concrete mixes, comprising RAC and NAC, is assessed by Braga et al (2017) to underline the role of factors such as cement quantity, superplasticizer, strength grade of concrete, and w/c. They observed that, better quality of coarse RCA, specified by higher density and low porosity or water absorption, may be used for lowering the negative environmental impact of concrete. Use of superplasticizer in small quantity may be recommended if it indirectly helps in reducing cement content in concrete without increasing the cost.

The role of various parameters such as functional unit, system boundary, allocation methods, and CO_2 uptake model on the life-cycle analysis (LCA) of RAC are explored previously (Zhang et al. 2019). To perform life-cycle analysis for RAC, different functional units (reference parameter to compare the input as well as environmental impact) are considered in the past as follows: (i) one cubic metre volume of concrete production, (ii) volume as well as strength that is one cubic metre volume of concrete having equal strength of NAC and RAC (Tošić et al. 2015; Serres et al. 2016), (iii) volume, strength as well as durability parameters that is for same 1 m^3 volume concrete of a specified strength grade with same service life (Knoeri et al. 2013), and (iv) reliability of strength, that is same coefficient of variation for RAC as well as NAC (Rodriguez-Nikl et al. 2012). Regarding the role of allocation methods on LCA results, Zhang et al (2019) reported that use of economic allocation method in the analysis can reduce energy consumption and emissions. Mass allocation can be preferable for by-products used as SCMs such as fly ash, slag, and silica fume in concrete (Seto et al. 2017).

The capacity of recycling plants and the distance of transportation from demolition sites to the recycling plant and construction sites need to be optimized to minimize energy consumption and CO_2 emissions (Coelho and de Brito 2013). The use of mobile recycling plants and linking life-cycle assessment with geospatial analysis to locate the plants can mitigate transportation-related impacts (Göswein et al. 2018). It has been found that higher cement content in concrete results in a greater global warming potential (GWP) regardless of aggregate types (Rosado et al. 2017). High strength RAC and expansion of functional unit may lead to increased CO_2 emissions due to higher cement requirement (Visintin et al. 2020).

4.2 Environmental Impact Assessment of RAC Containing SCMs

SCMs, such as fly ash, slags, silica fume, diatomite, etc., produced from other primary industrial processes, can share a portion of the environmental impacts associated with the primary process, when utilized as a constituent in concrete. For example, the environmental impact of electricity generation is assigned to fly ash preferably by the economic allocation method rather than the mass allocation method (Marinković et al. 2017). The material emission factor is lower in fly ash as compared to slag; the reason is lesser processing requirements for fly ash, and proximity of many coal-fired power stations reduces transportation-related distances (Gracia et al. 2014).

The environmental impacts of combined RCA and SCMs are assessed by Xing et al. (2022). They reported that greater the replacement of cement substituted SCMs and natural aggregate replaced recycled concrete aggregates, lower the associated carbon emissions. SCMs are largely beneficial in reducing global warming potential of concrete. The use of RCA can reduce the ozone layer depletion potential. When comparing fly ash, slag, and silica fume, it is shown that silica fume has the largest environmental burdens and is also the most sensitive, regardless of the allocation coefficients chosen.

4.2.1 *Effect of Fly Ash and Coarse RCA*

This section discusses cradle-to-gate life-cycle assessment performed by the authors using the SimaPro software tool for FARAC. The environmental impacts are primarily recorded in terms of global warming potential (GWP), abiotic depletion potential (ADP), photochemical oxidation potential, ozone layer depletion potential (ODP), and eutrophication for the concrete mix designed using PPM and the conventional Indian standard method. Fly ash produced in India contains higher carbon, greater variability, and mostly low-calcium types. Economic allocation coefficients, transportation distances, and recycling facility of coarse aggregates also differ in regional basis. Therefore, the authors assess the environmental impact of RAC and fly ash in the Indian context.

LCA of concrete was carried out for the raw materials production (quantities based on the mix design); energy and fuel used in the process (given in the LCI) and the associated transportation processes (included within the specified system boundary). Environmental impacts of NAC and FARAC were compared. Also, the contribution of each input materials on the output impact categories was assessed.

Another important aspect is the recycling process. Recycling processes for both these wastes are different based on their post-production utilization. Construction demolished wastes are generated at the end of use phase of existing concrete structures. Fly ash is produced as a by-product of the electricity generation. Recycling of CDW is termed as closed-loop (Marinković et al. 2017). It means the recycled products (in this case RCA) are used in the same concrete construction process from which CDW is produced. On the other hand, open-loop recycling means that the waste gets recycled into some other form to be used in another process or product formation. Fly ash collected from the electricity production process and utilized as a SCM in concrete indicates open-loop recycling. So, the impact of raw materials (input) and output emissions associated with the primary process of electricity generation needs to be allocated to fly ash in some proportion. Based on ISO 14040 recommendations, it is preferable to adopt system expansion for primary process instead of allocation. However, if system expansion is not possible, then one can go for more physical basis with mass allocation or economic allocation. Regardless of the unavoidable fluctuations in the market price associated with economic allocation, fly ash being a waste, should account for a smaller contribution of impact associated with the electricity generation (Marinković et al. 2017). Beneficial effect of fly ash utilization can be significantly witnessed through economic allocation (Chen et al. 2010). The authors implemented economic allocation method in the LCA of FARAC as discussed in the following sections.

4.2.1.1 Goal and Scope

The environmental impact of ten different concrete mixes, including low-volume fly ash (up to 30%) incorporated RAC and NAC, was assessed. The concrete mixes were proportioned using particle packing method and Indian standard method of mix designs. The goal of this study is to select the best possible alternatives among these concrete mixes to achieve lesser environmental burdens without compromising the compressive strength of concrete. The scope of the study is limited to the following aspects: (i) recycling process of construction demolition waste to produce coarse aggregate, (ii) extraction of natural coarse aggregate and sand, (iii) transportation of aggregates, (iv) production of cement, (vi) production of fly ash, and (vi) use of admixture.

The life-cycle assessment was carried out to produce 1 m^3 concrete which is considered as the functional unit (FU). FU is defined as a measure to relate both the input raw materials and processes to the output performance parameter such as compressive strength (Zhang et al. 2019). As mentioned before, the 28-day CS of NAC and RAC with fly ash were found equivalent which justifies the consideration of 1 m^3 as the functional unit for concrete. Raw materials' quantities calculated using the two different mix design methods are shown in Table 4.1.

4.2.1.2 System Boundaries

After deciding the FU, we need to identify the system boundary that includes all the energy and material flow processes to achieve the specified goal (Tillman et al. 1994). Cradle-to-gate analysis was performed for the system boundary as depicted in Fig. 4.1. This system boundary includes the phases like raw material processing,

Table 4.1 Mix proportions and compressive strength of NAC and RAC mixes

Mix type	Method	Cement kg/m^3	Fly ash kg/m^3	NCA kg/m^3	Sand kg/m^3	RCA kg/m^3	CS (28 days)
NAC	PPM	410	–	1143	762	–	42.75
NAC	IS code	438	–	1242	610	–	45.37
NFA20	PPM	361	90	1143	762	–	44.51
NFA30	PPM	316	135	1143	762	–	42.24
NFA20	IS	372	93	1242	610	–	46.07
NFA30	IS	325	139	1242	610	–	42.18
RFA20	PPM	361	90	–	659	1007	43.22
RFA30	PPM	316	135	–	659	1007	41.82
RFA20	IS code	372	93	–	600	1079	44.66
RFA30	IS code	325	139	–	600	1079	41.29

transportation of CDW to the recycling plant along with the generation of RCA, production of cement and admixture in factory, extraction or processing of natural coarse aggregates and sand, transportation of all the constituent materials and water from their respective production factory or source to the concrete batching plant and production of concrete. As the concrete mixes are designed for equivalent compressive strength, no other differences will appear during the life-cycle phases up to concrete manufacturing. Other processes like transportation of concrete to the site, construction of structures, service, and demolition (end-of-life) phases may vary based on the use of the structures (Jiménez et al. 2015); hence, these phases were kept outside the system boundary.

4.2.1.3 Preparation of Life-Cycle Inventory (LCI)

Preparation of LCI includes collection of input data regarding energy and material flow in the processes involved and the associated output emissions. All the data regarding raw material production, processing, and transportation correspond to the Indian scenario and collected from the local plants for recycling construction wastes, cement, and fly ash. Data related to the production of 1 tonne of NCA and recycling of RCA were collected from (Pradhan et al. 2019), a study based on concrete recycling plant located in New Delhi, India. The production of NCA involves excavating basalt rock from a quarry using explosives. Excavators dig out the rock, load it on to lorries, and transport it to the crushing plant having production capacity of 1500 tonne/day. The fuel consumption rate was considered as 5 litre per hour in the process of excavation and 26.5 litre per hour for transportation. The rocks are crushed using jaw crushers and impact crushers to break them down into smaller sizes. This crushing process consumed primary energy of 700 kW. After proper sorting, the coarse aggregates are separated into different sizes, with a greater emphasis on

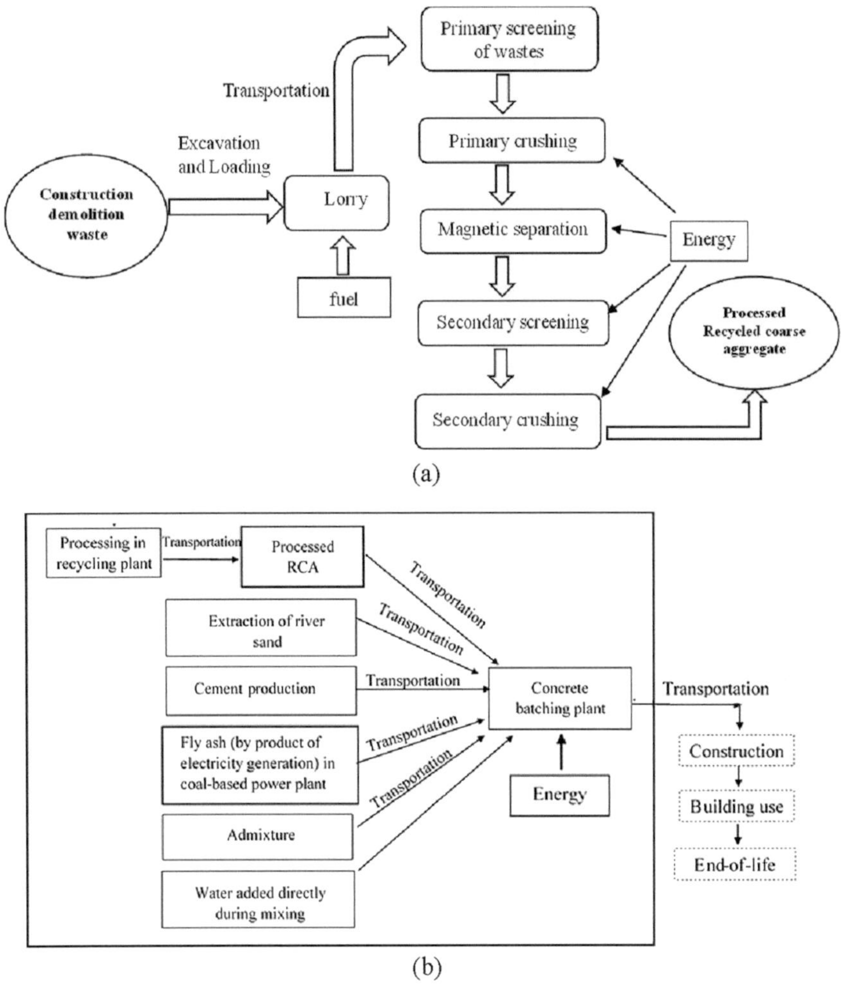

Fig. 4.1 System boundary for FARAC **a** processing and production of RCA, **b** production of concrete

producing aggregates of 20 mm down size (which are mostly used in construction), including sizes from 10 mm up to 6.3 mm or 4.75 mm. Electricity and fuel consumption in operation of machines for NCA production were considered as 2.72 kWh and 8.79 MJ, respectively.

The production of RCA involves transporting construction demolition waste to a recycling plant, where other wastes such as plaster, small bricks, or hardened cement paste are manually removed (Hansen 1985; Coelho and debrito 2013). Then, the larger blocks are crushed into smaller sizes. Dust emissions in the plant are controlled by using water sprinklers and suction bags. For undergoing these steps in a recycling plant of capacity 280 tonne/day, approximately 122.8 kW power is required. Different-sized aggregates are sorted and stored in open-air piles. Electricity consumption in processing 1 tonne of recycled coarse aggregates was considered as 2.12 kWh. Quantity of fuel required was 7.92 MJ. Transportation distances related to recycling of RCA is according to the Indian conditions where there are very few recycling plants. A 16–32-tonne capacity lorry was used to transport basalt from the quarry to NCA production plant (5 km) and CDW from demolition site to the recycling plant (35 km). A 7.5–16-tonne capacity lorry was used for other related transportation of raw materials such as cement (from the factory), sand (from the river), and admixture (from the production plant) up to the concrete batching plant. The respective distances (in km) for these transports were taken as 70 km, 30 km, and 100 km, respectively (Pradhan et al. 2019).

Fly ash, used in this study as a partial cement substitute, was obtained as a by-product from the coal-based power plant. Due to the abundance of such plants in India, the average distance over which fly ash is transported for concrete production was taken as 50 km. As mentioned earlier, raw materials like hard coal (prevalent in Indian condition) and emissions of different harmful gaseous particulates associated with electricity generation were allocated to fly ash using the economic allocation method. Greater quantities of ash content, lower calorific values, and less-efficient combustion technologies of the India-based thermal power plants attribute for larger emissions to the environment (Chakraborty et al. 2008). Data collected from the local thermal power plants in West Bengal (India) indicate that the production of 1 kwh of electricity requires 0.62 kg of hard coal and generates a by-product of 0.1438 kg of fly ash. In the Indian market, the price of fly ash is generally around Rs. 2500 (30.33 USD) per tonne while electricity is priced as Rs. 6/kWh. The economic allocation coefficient ($C_{e,flyash}$) was calculated using the formula given by (Chen et al. 2010) as shown in Eq. 4.1.

$$C_{e,fly\ ash} = \frac{(Rs.m)_{byproduct}}{(Rs.m)_{byproduct} + (Rs.m)_{primary\ product}} \tag{4.1}$$

$$C_{e,fly\ ash} = \frac{0.1438kg \times 2.5Rs/kg}{\left(1kwh \times 6\ Rs + 0.1438\ kg \times 2.5Rs/kg\right)} = 0.0565 \tag{4.2}$$

Here, Rs signifies price (in rupees) per unit of the product and m stands for total mass of the material produced in the process. $C_{e,fly\ ash}$ is used as a multiplicative factor when considering the input quantity of raw materials (hard coal) and emission data associated with electricity generation (Marinković et al. 2017). This factor helps determine the contribution of fly ash in the overall calculation. The emissions of major gases such as CO_2, CO, SO_2, and NO from coal-based thermal power plants in India were obtained from Chakraborty et al. (2008). To produce 1 kWh electricity, suggested average emission coefficients were 8.696 gm, 3.393 gm, 2.42 gm, and 0.998 kg, respectively, for SO_2, CO, NO, and CO_2.

The life-cycle assessment (LCA) was conducted using SimaPro software package, version 9. A life-cycle inventory was prepared for the raw materials involved in concrete production using the available data as mentioned above. The data which are unavailable for the Indian condition were considered from the Ecoinvent 3.5 database as mentioned in the following modules: 'cement, Portland {ROW}' for cement, 'sand {ROW}' for sand, 'water, deionised from tap water, at user {ROW}' for water, 'diesel, burned in building machine (GLO)' for diesel and 'electricity, medium voltage {IN}' for electricity. Admixture used in the production of RAC was not available in the ecoinvent module. Hence, these data were taken from LCI of polycarboxylate-based admixture recommended by the European Federation of Concrete Admixture Associations 2015. By using these LCI data, life-cycle assessment was carried out. Different impact categories were assessed, and the results were compared for NAC, NFA20, NFA30, RFA20, and RFA30.

This step estimated potential environmental impacts associated with different concrete in terms of impact categories such as abiotic depletion (fossil fuel), global warming potential (GWP), ozone layer depletion potential (ODP), human toxicity, fresh and marine water ecotoxicity, terrestrial ecotoxicity, photochemical oxidation (PCO), eutrophication, and acidification. CML baseline mid-point method was considered in the present study for the impact assessment. Based on the available emission data, fly ash contributed significantly to four impact categories such as PCO, GWP, eutrophication, and acidification. Effect of other input raw materials, transportation, diesel, and electricity were also plotted for these categories separately. NAC with and without fly ash and RAC with fly ash concrete were compared for the considered environmental impacts.

Contributions of input energy and materials are represented in Fig. 4.2. Cement was observed as the highest contributing factor to adverse environmental impact followed by the effect of transportation and fly ash. Cement contributed for about 92% in NAC and 76% in RFA30 (fly ash accounted for about 13%) as far as GWP is concerned. Use of fly ash as cement substitute in concrete helps in reduction of GWP as it facilitates lowering of cement content (Teixeira et al. 2016; Panesar et al. 2019). PCO was affected by about 82% in NAC and 65% in RFA30 with similar fly ash effect as in GWP. Cement accounted for 79% and 69% (fly ash effect = 7.6%) in case of acidification respectively for NAC and RFA30. Similarly, 73% to 60% (fly ash = 12%) was affected by cement respectively in NAC and RFA30 in the case of eutrophication. PPM mix design reduced the effect of cement by around 1.5–2%. Besides cement, transportation attributed to about 6, 12, 12, and 16% in NAC for GWP, PCO, acidification, and eutrophication respectively. On the other hand, RFA30 accounted for 8% in GWP, 16%, 18%, and 21% in PCO, acidification, and eutrophication respectively as far as contribution of transportation was concerned. The order of impact observed was as follows: cement > transportation > fly ash > sand > electricity.

4.2.1.4 Life-Cycle Assessment

Figure 4.3 illustrates environmental impact assessment of different concretes underlining the effects of PPM and IS mix design. The environmental impact is assessed in terms of different categories such as abiotic depletion (AD), global warming potential

Fig. 4.2 Contribution on **a** GWP, **b** POCP, **c** acidification, **d** eutrophication for different concrete

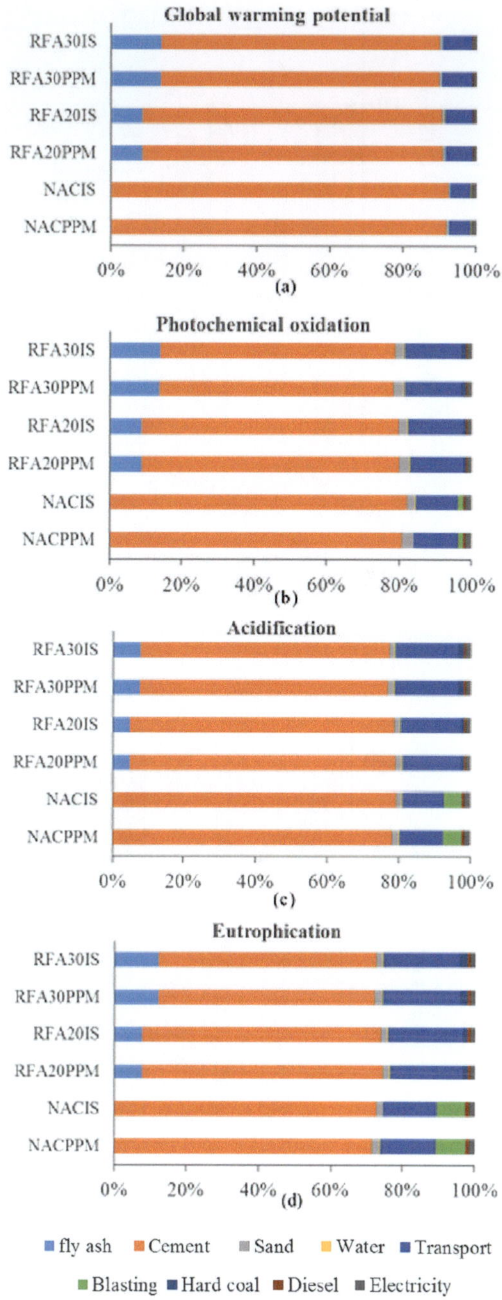

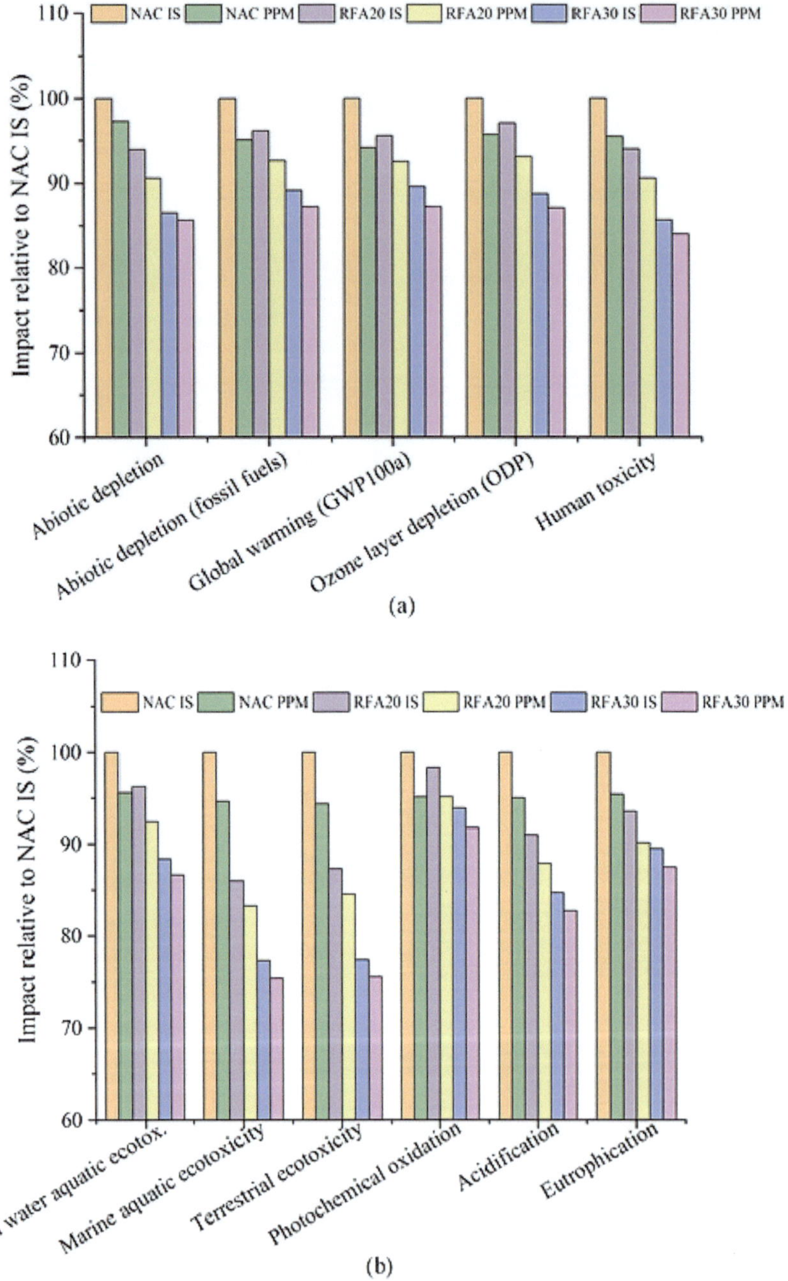

Fig. 4.3 Comparison of different concrete based on mix design for adverse environmental impact

(GWP), photochemical oxidation (PCO), acidification, eutrophication, ecotoxicity, and ozone layer depletion (OD). Values shown in the figure represent impact relative to NAC proportioned using IS method. NAC prepared using conventional IS method shows the highest adverse environmental impact. The effectiveness of PPM can be observed by comparing NACPPM, RFA20PPM, and RFA30 PPM with NACIS, RFA20IS, and RFA30IS, respectively. As compared to IS mix design, concrete proportioned using PPM showed about 2–5% reduction in the negative effects. This is due to the use of lesser cementitious materials in PPM mix design. The combined influence of 100% RAC with fly ash and PPM can be inferred from the comparison of RFA20PPM and RFA30PPM with the control concrete denoted as NACIS. The impact such as AD, GWP, OD, PCO, and acidification reduced by 9, 7.5, 7, 5, and 12% in RFA20PPM. Similarly, RFA30PPM showed about 14, 12.8, 13, 8, and 17% lesser values for the above-mentioned parameters. Kurda et al. (2018) also observed positive effects considering environmental and cost aspects of concrete for 100% RCA, 1% superplasticizer, and 30% fly ash satisfying strength and durability. In a nutshell, the combined influence of up to 30% fly ash, 100% RCA, and PPM can be helpful in reducing the net adverse environmental impact of concrete without compromising strength of concrete.

4.3 Summary

- Effects of emissions, and raw material associated with electricity generation, were allocated to fly ash through economic allocation method considering open loop recycling. Cement contributed for about 92% in NAC and 76% in RFA30 as far as GWP is concerned. Contribution of raw materials towards different environmental impact categories was obtained in the following order: cement > transportation > fly ash > sand.
- Natural aggregate concrete produced with conventional method of mix design results in the maximum adverse environmental impact. As compared to IS mix design, concrete proportioned using PPM showed about 2–5% reduction in the negative effects. This is due to the use of lesser cementitious materials in PPM mix design.
- The impact categories such as AD, GWP, OD, PCO, and acidification reduced by 9, 7.5, 7, 5, and 12% in RFA20PPM as compared to NAC proportioned with IS mix design method. Similarly, RFA30PPM showed about 14, 12.8, 13, 8, and 17% lesser values for the above-mentioned parameters.
- Combined influence of fly ash as partial cement substitute, 100% recycled coarse aggregates, and particle packing method of mix design is found highly beneficial in minimizing adverse environmental impact of concrete without affecting strength requirements.

References

Braga AM, Silvestre JD, de Brito J (2017) Compared environmental and economic impact from cradle to gate of concrete with natural and recycled coarse aggregates. J Clean Prod 162:529–543

Chakraborty N, Mukherjee I, Santra AK, Chowdhury S, Chakraborty S, Bhattacharya S, Sharma C (2008) Measurement of CO2, CO, SO2, and NO emissions from coal-based thermal power plants in India. Atmos Environ 42:1073–1082

Chen C, Habert G, Bouzidi Y, Jullien A, Ventura A (2010) LCA allocation procedure used as an incitative method for waste recycling: an application to mineral additions in concrete. Resour Conserv Recycl 54(12):1231–1240

Coelho A, de Brito J (2013) Environmental analysis of a construction and demolition waste recycling plant in Portugal-Part I: energy consumption and CO2 emissions. Waste Manage 33(5):1258–1267

García-Segura T, Yepes V, Alcalá J (2014) Life cycle greenhouse gas emissions of blended cement concrete including carbonation and durability. Int J Life Cycle Assess 19:3–12

Göswein V, Gonçalves AB, Silvestre JD, Freire F, Habert G, Kurda R (2018) Transportation matters–Does it? GIS-based comparative environmental assessment of concrete mixes with cement, fly ash, natural and recycled aggregates. Resour Conserv Recycl 137:1–10

Hansen TC (1985) Recycled aggregates and recycled aggregate concrete second state-of-the-art report developments 1945–1985. Mater Struct 19:201–246

ISO (2006) Environmental management—life cycle assessment, set of international standards: ISO 14040e14043. International Organization Standardization, Geneva

Jiménez C, Barra M, Josa A, Valls S (2015) LCA of recycled and conventional concretes designed using the equivalent mortar volume and classic methods. Constr Build Mater 84:245–252

Knoeri C, Sanyé-Mengual E, Althaus HJ (2013) Comparative LCA of recycled and conventional concrete for structural applications. Int J Life Cycle Assess 18:909–918

Kurda R, Silvestre JD, de Brito J (2018) Life cycle assessment of concrete made with high volume of recycled concrete aggregates and fly ash. Resour Conserv Recycl 139:407–417

Marinković S, Radonjanin V, Malešev M, Ignjatović I (2010) Comparative environmental assessment of natural and recycled aggregate concrete. Waste Manage 30(11):2255–2264

Marinković S, Dragaš J, Ignjatović I, Tošić N (2017) Environmental assessment of green concretes for structural use. J Clean Prod 154:633–649

Panesar DK, Kanraj D, Abualrous Y (2019) Effect of transportation of fly ash: Life cycle assessment and life cycle cost analysis of concrete. Cement Concr Compos 99:214–224

Pradhan S, Tiwari BR, Kumar S, Barai SV (2019) Comparative LCA of recycled and natural aggregate concrete using Particle Packing Method and conventional method of design mix. J Clean Prod 228:679–691

Rodriguez-Nikl T, Christiansen JW, Walters K (2012) Reliability-based life cycle assessment of green concrete structures. Special Publication 289:1–13

Rosado LP, Vitale P, Penteado CSG, Arena U (2017) Life cycle assessment of natural and mixed recycled aggregate production in Brazil. J Clean Prod 151:634–642

Serres N, Braymand S, Feugeas F (2016) Environmental evaluation of concrete made from recycled concrete aggregate implementing life cycle assessment. J Build Eng 5:24–33

Seto KE, Churchill CJ, Panesar DK (2017) Influence of fly ash allocation approaches on the life cycle assessment of cement-based materials. J Clean Prod 157:65–75

Teixeira ER, Mateus R, Camoes AF, Bragança L, Branco FG (2016) Comparative environmental life-cycle analysis of concretes using biomass and coal fly ashes as partial cement replacement material. J Clean Prod 112:2221–2230

Tillman AM, Ekvall T, Baumann H, Rydberg T (1994) Choice of system boundaries in life cycle assessment. J Clean Prod 2(1):21–29

Tošić N, Marinković S, Dašić T, Stanić M (2015) Multicriteria optimization of natural and recycled aggregate concrete for structural use. J Clean Prod 87:766–776

Visintin P, Xie T, Bennett B (2020) A large-scale life-cycle assessment of recycled aggregate concrete: The influence of functional unit, emissions allocation and carbon dioxide uptake. J Clean Prod 248:119243
Xing W, Tam VW, Le KN, Hao JL, Wang J (2022) Life cycle assessment of recycled aggregate concrete on its environmental impacts: a critical review. Constr Build Mater 317:125950
Zhang Y, Luo W, Wang J, Wang Y, Xu Y, Xiao J (2019) A review of life cycle assessment of recycled aggregate concrete. Constr Build Mater 209:115–125

Part II
Properties of Green Concrete

Chapter 5
Fresh Concrete Properties

Preview

Concrete constituted of coarse recycled concrete aggregate and supplementary cementitious materials must display technical properties similar to or within an acceptable range of deviation when compared to the existing conventional concrete. This enables reliable adoption of the new materials as alternative concrete constituents in practice. This chapter explores the material level concrete properties in the fresh state focusing on the role of coarse recycled concrete aggregates (coarse RCA) and challenges associated with it. Then the synergistic effects of SCMs and coarse RCA in concrete are explored. Fresh concrete properties focused here are workability, segregation, and bleeding, and setting time. The roles of SCMs discussed in this chapter are limited to the widely available materials such as fly ash, slag, and silica fume.

5.1 Workability

5.1.1 Role of Recycled Aggregates

The physical characteristics of coarse aggregates, such as shape and surface texture, can have an impact on the fresh properties of concrete (Neville 2002). Coarse aggregates that have elongated or flaky particle shapes tend to increase the water demand. So, equidimensional coarse aggregates are generally preferred in concrete. The surface texture of aggregates impacts the bond between aggregate and paste, thereby influencing the water demand of concrete mix.

With this background on the key traits of natural coarse aggregate that affect the properties of fresh concrete, we will now discuss the impacts of replacing coarse RCA on the workability of concrete. The application of coarse RCA reduces workability

S. Sunayana and S. V. Barai, *Green Concrete*, Springer Transactions in Civil and Environmental Engineering, https://doi.org/10.1007/978-981-96-6982-0_5

of concrete due to the presence of low density adhered residual mortar layer (Hansen and Narud 1983). This leads to higher water absorption. Furthermore, increasing the proportion of coarse RCA results in a more significant reduction in slump. The non-uniform shape and rough surface texture of coarse RCA increase internal friction and decrease workability (Lima et al. 2013). These factors like shape, size, and surface texture of coarse aggregate are also regulated by the crushing procedure which in turn affect the workability (Ibrahim et al. 2023). Coarse RCA produced by the process of both primary and secondary crushing resulted in relatively rounded aggregate which make less workable concrete mix as compared to angular aggregate processed from only the primary crushing method.

The moisture state of coarse RCA may be another important parameter affecting the workability of concrete incorporating it. Poon et al (2004) conducted a study to examine how the workability of concrete is affected by using coarse RCA under different moisture conditions such as oven-dry, air-dry, and surface-saturated dry state. They reported that, initial slump of concrete is higher when coarse RCA in the oven-dry state is used; the reason is greater effective w/c ratio. However, after 15–30 min of mixing, all concrete mix showed comparable workability when the additional quantity of water added to adjust the higher water absorption of coarse RCA was completely absorbed by the mix. This pattern of slump loss based on different moisture state of coarse RCA-based concrete is depicted in Fig. 5.1. This figure is adapted from Silva et al. (2018).

Following methods are recommended in the literature to counteract the adverse impact of coarse RCA on the workability of concrete: (1) requirement of additional amount of water (about 15%) or water compensation method (Frondistou-Yannas 1977; Malhotra 1978; Rasheeduzzafar and Khan 1984; Topçu 1997; Ferreira et al. 2011); (2) incorporating extra cement or reducing w/c (Sagoe-Crentsil et al. (2001); González-Fonteboa and Abella 2008); (3) pre-saturation of RCA (Ferreira et al. 2011); and (4) use of water-reducing admixture (Pedro et al. 2017). The water compensation method involves adding extra water to the concrete during mixing to account for the higher water absorption of coarse RCA. Accurate implementation of this technique also necessitates the precise assessment of rate of water absorption of coarse RCA. The pre-saturation process involves immersing coarse RCA in water for a duration of 24 h, followed by air-drying, to achieve a surface-saturated dry state prior to their use in the concrete mix (Silva et al. 2018). According to Beltran et al. (2014), increasing the amount of cement by 12% can result in a workable concrete mix when 100% coarse RCA is utilized. Regardless of these methods, lower slump values were observed for recycled concrete produced with equivalent w/c as that of NAC (Etxeberria et al. 2007; Yang et al. 2011; Viera et al. 2011; Sagoe-Crentsil et al. 2001; González -Fonteboa et al. 2011). In addition to the methods mentioned above, the adverse impacts of coarse RCA on workability of concrete can be mitigated by incorporating supplementary cementitious materials as a partial replacement for cement. The following section focuses on the impact of frequently employed used SCMs on the workability of concrete made with coarse RCA.

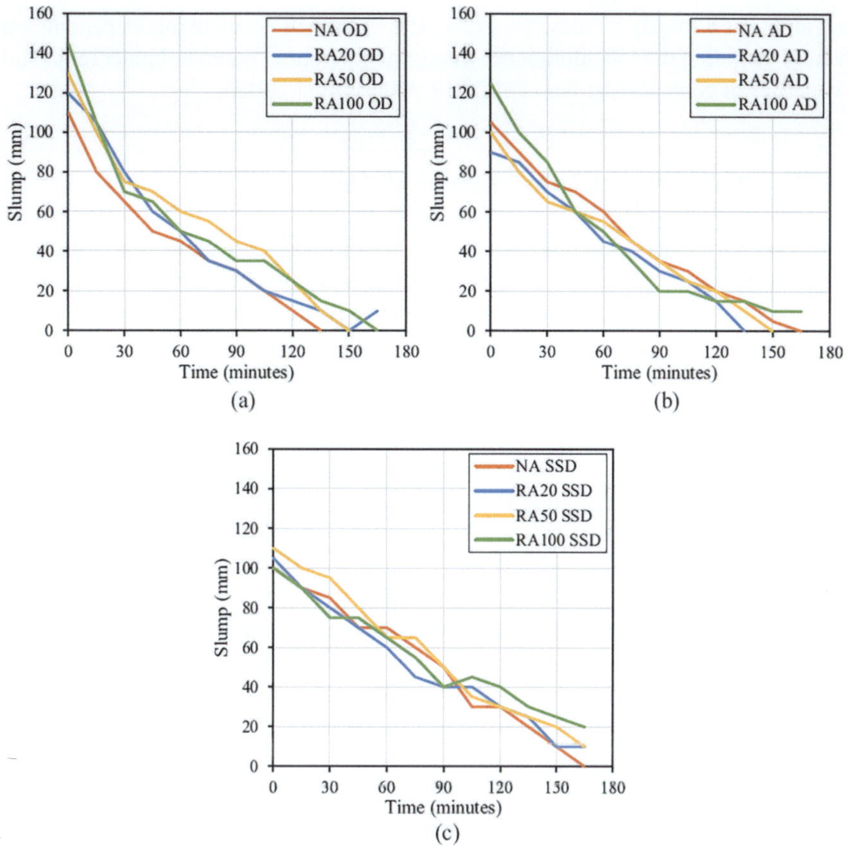

Fig. 5.1 Slump loss of RAC over time for concrete mixes with varying replacement level **a** oven-dry, **b** air-dry, **c** surface-saturated dry state (adapted from Silva et al. 2018)

5.1.2 Role of Cementitious Materials

In the natural aggregate concrete, incorporating fly ash as a substitute for cement (up to 20%) has been proven advantageous in enhancing workability or decreasing water requirement, as long as the fly ash particles are spherical, finer than cement, and have low carbon content (Giergiczny 2019). The spherical fly ash grains of size 1–100 μm, constituted of glassy phase can decrease the amount of water needed in a concrete mix (Ramezanianpour 2014). Equal volume of cement substituted fly ash results in higher paste volume due to lesser specific gravity and higher fineness. Besides workability, use of fly ash is also beneficial in reducing bleeding and segregation of concrete mix (Sideris et al. 2018). As grain morphology of fly ash affects workability, high calcium fly ash may show increased water demand if more active compounds like free CaO are present (Giergiczny 2019). The grain filling effect and glassy phase of granulated blast furnace slags (GBFS) also improve workability of the concrete. However, fly ash and GBFS mostly increase the initial setting time of concrete mix depending on factors including the substitution percentage, temperature, water-to-binder ratio, and properties of cement (Matthes et al. 2018; Sideris et al. 2018).

5.1.2.1 Role of Fly Ash on Workability of RAC

In this section, let us discuss the effects of SCMs on concrete made of coarse RCA. In their study, Poon et al. (2004) observed that the inclusion of fly ash in RAC resulted in an increase in slump and a decrease in the rate of slump loss over time. Previous literature (Li et al. (2009); Kou and Poon 2012; Du et al. 2018) has documented the increase in workability due to the lubrication effect of spherical pozzolanic materials, particularly fly ash. Superplasticizer in a moderate percentage of about 0.37–0.42% can be used to achieve desired workability (100 ± 20) in concrete constituting 100% coarse RCA and upto 30% fly ash (Sunayana and Barai 2017). Additionally, the fresh state of FARAC exhibited a higher setting time and a relatively smoother surface finish.

5.1.2.2 Role of Slag and Silica Fume on Workability of RAC

Ground granulated blast furnace slag (GGBFS) used as 50% of cement replacement in recycled concrete showed improvement in the workability (Berndt, 2009). This is also supported by Kou et al. (2011), who observed an enhanced workability of concrete due to the lubrication effect of GGBFS regardless of the presence of coarse RCA. On the other hand, workability of 100% RAC comprising both coarse and fine recycled concrete aggregate is negatively affected (Moghadam et al. 2021). A reduction of 10, 21, and 31% in the slump of concrete was observed when 15, 30, and 45% GGBFS used as cement substitutes in RAC. This adverse effect on workability was attributed to irregular shape and rough texture of GGBFS particles. The addition of 10% silica fume or 15% metakaolin to concrete mixes has a minimal impact on the decrease in slump, regardless of the amount of coarse RCA in concrete (Kou et al. 2011).

5.2 Fresh Density

Mass per unit volume of concrete in the initial fresh state prior to setting is referred to as fresh density. This parameter can also affect the properties of concrete in its hardened state. Lesser porosity and entrapped air content in concrete imply higher fresh density. Thus, coarse RCA and other SCMs incorporated in concrete are expected to have an impact on its fresh density (Ibrahim et al. 2023). The roles of factors such as shape, surface texture, and percentage replacement of coarse RCA are important in this regard. The presence of the old adhered mortar layer on RCA surface lowers its fresh density as compared to that of NCA. Thus, using coarse RCA might lead to a decrease in the density of fresh concrete. Silva et al. (2018) reported that inclusion of 100% coarse RCA resulted in a 5% reduced density. The inferiority of lower fresh density may be mitigated by using coarse RCA which are produced through both primary and secondary crushing methods (Lavado et al. 2020).

5.3 Air Content

The air content of RAC is influenced by its surface roughness, shape, and the proportion of coarse RCA used (Lavado et al. 2020). The coarse RCA produced just through primary crushing exhibit a 14% increase in air content due to their rough surface texture and angular shape. Conversely, when both primary and secondary crushing methods are employed, the resulting coarse RCA showed a 10% decrease in air content compared to that of NAC. Increasing the coarse RCA replacement leads to a higher air content. The weak mortar layer attached to the surface of the RCA contributes to the increased air content (Wang et al. 2021).

5.4 Bleeding and Segregation

Immediately after the concrete mix is prepared, a layer of free water may get accumulated on the surface, a phenomenon known as bleeding of concrete. This leads to the formation of non-homogenous mix. According to the study of Dhir et al. (1999) and Silva et al. (2018), higher coarse RCA replacement percentages were associated with higher cumulative bleeding up to approximately 45 min after mixing. However, the bleeding water gradually decreased after that and, two hours later, the extra bleed water in 100% coarse RCA and 50% fine RCA replaced concrete was 17% lower. In a study conducted by Poon et al. (2007), the concrete comprising about 100% coarse RCA resulted in 25% higher bleeding water.

5.5 Summary

In this chapter, the fresh concrete properties such as workability, fresh density, air content, bleeding and segregation are discussed for RAC with and without SCMs. The shape, surface texture and percentage substitution of coarse RCA are found to be among the important factors affecting fresh properties of concrete. The crushing method, where coarse RCA is generated by both primary and secondary crushing, produces aggregates causing lesser inferiorities as far as the fresh concrete properties are concerned. The spherical pozzolanic particles of fly ash, slag, and silica fume contribute to improvement in workability. In 100% coarse RCA replaced concrete comprising up to 30% fly ash as cement substitute, superplasticizer (upto about 0.4% of total cementitious materials) may be used for achieving desired workability. Fresh density and air content in RAC may be higher due to the weak layer of adhered mortar content.

References

Beltrán MG, Barbudo A, Agrela F, Galvín AP, Jiménez JR (2014) Effect of cement addition on the properties of recycled concretes to reach control concretes strengths. J Clean Prod 79:124–133

Berndt ML (2009) Properties of sustainable concrete containing fly ash, slag and recycled concrete aggregate. Constr Build Mater 23(7):2606–2613

Dhir RK, Limbachiya MC, Leelawat T (1999) Suitability of recycled concrete aggregate for use in BS 5328 designated mixes. Proc Inst Civil Eng Struct Build 134(3)

Du X, Tao W, Fu X, Liu M (2018) Workability and mechanical property of low grade pumping concrete modified by recycled coarse aggregate. IOP Conf Ser: Mater Sci Eng 397(1)

Etxeberria M, Vázquez E, Marí A, Barra M (2007) Influence of amount of recycled coarse aggregates and production process on properties of recycled aggregate concrete. Cem Concr Res 37(5):735–742

Ferreira L, De Brito J, Barra M (2011) Influence of pre-saturation of recycled coarse concrete aggregates on structural concretes mechanical and durability properties. Mag Concr Res 63(8):617–627

Frondistou-Yannas S (1977) Waste concrete as aggregate for new concrete. ACI J Proc 74(8):373–374

Giergiczny Z (2019) Fly ash and slag. Cem Concr Res 124:105826

González-Fonteboa B, Martínez-Abella F (2008) Concretes with aggregates from demolition waste and silica fume. Materials and mechanical properties. Build Environ 43(4):429–437

González -Fonteboa B, Martinez-Abella F, Eiras-Lopez J, Seara-Paz S (2011) Effect of recycled coarse aggregate on damage of recycled concrete. Mater Struct 44:1759–1771

Hansen TC, Narud H (1983) Strength of recycled concrete made from crushed concrete coarse aggregate. Concr Int 5(1):79–83

Ibrahim M, Alimi W, Assaggaf R, Salami BA, Oladapo EA (2023) An overview of factors influencing the properties of concrete incorporating construction and demolition wastes. Constr Build Mater 367:130307

Kou SC, Poon CS (2012) Enhancing the durability properties of concrete prepared with coarse recycled aggregate. Constr Build Mater 35:69–76

Kou SC, Poon CS, Agrela F (2011) Comparisons of natural and recycled aggregate concretes prepared with the addition of different mineral admixtures. Cement Concr Compos 33(8):788–795

Lavado J, Bogas J, De Brito J, Hawreen A (2020) Fresh properties of recycled aggregate concrete. Constr Build Mater 233:117322

Li J, Xiao H, Zhou Y (2009) Influence of coating recycled aggregate surface with pozzolanic powder on properties of recycled aggregate concrete. Constr Build Mater 23(3):1287–1291

Lima C, Caggiano A, Faella C, Martinelli E, Pepe M, Realfonzo R (2013) Physical properties and mechanical behaviour of concrete made with recycled aggregates and fly ash. Constr Build Mater 47:547–559

Malhotra VM (1978) Use of recycled concrete as new aggregate. In: Symposium on energy and resource conservation in the concrete industry, CANMET report no. 76–8, CANMET, Ottawa, Canada

Matthes W, Vollpracht A, Villagrán Y, Kamali-Bernard S, Hooton D, Gruyaert E, De Belie N (2018) Ground granulated blast-furnace slag. Properties of fresh and hardened concrete containing

supplementary cementitious materials: state-of-the-art report of the RILEM technical committee 238-SCM, Working Group 4, pp 1–53
Moghadam AS, Omidinasab F, Goodarzi SM (2021) Characterization of concrete containing RCA and GGBFS: mechanical, microstructural and environmental properties. Constr Build Mater 289:123134
Neville AM (2002) Properties of concrete, vol 4. Longman, London
Pedro D, De Brito J, Evangelista L (2017) Evaluation of high-performance concrete with recycled aggregates: use of densified silica fume as cement replacement. Constr Build Mater 147:803–814
Poon CS, Shui ZH, Lam L, Fok H, Kou SC (2004) Influence of moisture states of natural and recycled aggregates on the slump and compressive strength of concrete. Cem Concr Res 34(1):31–36
Poon CS, Kou SC, Lam L (2007) Influence of recycled aggregate on slump and bleeding of fresh concrete. Mater Struct 40(9):981–986
Ramezanianpour AA (2014) Cement replacement materials. Springer Geochem Mineral 10:978–983
Rasheeduzzafar AK, Khan A (1984) Recycled concrete-a source of new aggregate. Cement Concrete Aggregate 6(1):17–27
Sagoe-Crentsil KK, Brown T, Taylor AH (2001) Performance of concrete made with commercially produced coarse recycled concrete aggregate. Cem Concr Res 31(5):707–712
Sideris K, Justnes H, Soutsos M, Sui T (2018) Fly ash. Properties of fresh and hardened concrete containing supplementary cementitious materials: state-of-the-art report of the RILEM technical Committee 238-SCM, working group 4, pp 55–98
Silva RV, De Brito J, Dhir RK (2018) Fresh-state performance of recycled aggregate concrete: a review. Constr Build Mater 178:19–31
Sunayana S, Barai SV (2017) Recycled aggregate concrete incorporating fly ash: comparative study on particle packing and conventional method. Constr Build Mater 156:376–386
Topçu IB (1997) Physical and mechanical properties of concretes produced with waste concrete. Cem Concr Res 27(12):1817–1823
Vieira JPB, Correia JR, De Brito J (2011) Post-fire residual mechanical properties of concrete made with recycled concrete coarse aggregates. Cem Concr Res 41(5):533–541
Wang B, Yan L, Fu Q, Kasal B (2021) A comprehensive review on recycled aggregate and recycled aggregate concrete. Resour Conserv Recycl 171:105565
Yang J, Du Q, Bao Y (2011) Concrete with recycled concrete aggregate and crushed clay bricks. Constr Build Mater 25(4):1935–1945

Chapter 6
Hardened Concrete Properties

Preview

Concrete made with coarse recycled concrete aggregate (RCA) and supplementary cementitious materials (SCMs) must display technical properties similar to or in an acceptable range of deviation as compared to the existing conventional concrete. This enables reliable adoption of coarse RCA and SCMs as alternative concrete constituents in practice. In the previous chapter, we have discussed the effect of SCMs on fresh properties of recycled aggregate concrete. This chapter discusses the hardened concrete properties of concrete comprising of coarse RCA and challenges associated with it. Then the synergistic effects of SCMs and coarse RCA in concrete are explored. The mechanical properties focused here include compressive strength, tensile strength, and modulus of elasticity. The roles of SCMs discussed in this chapter are limited to the widely available fly ash, slag, and silica fume.

Introduction of SCMs also has an influence on hardened concrete properties, and the degree depends on percentage of replacement, physical properties like fineness, chemical and mineralogical composition, chemical reactivity, and curing age (Giergiczny 2019). In this chapter, let us discuss first, the mechanical properties of concrete comprising of coarse RCA and then the effect of SCMs on coarse RCA-based concrete, specifically focusing on the influence of fly ash on 100% coarse RCA-based concrete.

6.1 Effect of Coarse RCA on Compressive Strength (CS) and Tensile Strength of Concrete

The compressive strength of normal concrete relies mostly on the properties of interfacial transition zone (ITZ) which is regarded as the weakest link. As mentioned before, the main difference between the ITZ of RAC and that of NAC is the formation

S. Sunayana and S. V. Barai, *Green Concrete*, Springer Transactions in Civil and Environmental Engineering, https://doi.org/10.1007/978-981-96-6982-0_6

of two separate ITZs in the earlier. One transition zone is formed between the parent aggregate and old mortar, and another one is created between the old attached mortar and fresh paste matrix. Consequently, the utilization of coarse RCA in concrete has an impact on the characteristics of the ITZ, which subsequently affects the mechanical properties of concrete in the hardened state.

Key parameters affecting the compressive strength of coarse RCA-based concrete can be identified as replacement percentage, quality of RCA (defined by water absorption and density), parent concrete quality, crushing methods, mix design technique, and mixing method. The higher water absorption of RCA and lower availability of free water for hydration result in lesser compressive strength of recycled concrete (Gonzalez-Fonteboa and Abella 2005). According to Silva et al. (2015), replacing up to 30% of coarse recycled concrete aggregate (RCA) does not have any detrimental impact on the compressive strength of concrete. Replacement of coarse RCA up to 30% does not cause any negative influence on compressive strength of concrete, beyond which CS reduces. When natural coarse aggregate is completely replaced, the compressive strength of concrete decreases by 25–30% according to several studies (Ravindrarajah and Tam, 1985; Xiao et al. 2005; Etxeberria et al. 2007; Rahal, 2007; Saffiuddin et al. 2011; Butler et al. 2013). Use of 5–10% extra cement may compensate for the compressive strength loss in 100% coarse RCA-based concrete (Etxeberria et al. 2007). As discussed previously, the water compensation approach, which involves adding additional water to the concrete mix to account for the higher water absorption of coarse RCA, or using pre-saturated RCA, has been proven to be advantageous in obtaining similar compressive strength of RAC and NAC (Silva et al. 2015). Concrete made using high-quality RCA, which has a water absorption of less than 3% and an oven-dry density of more than 2.5 g/cm^3, results in a 9% decrease in CS compared to NAC. On the other hand, using low-quality RCA with a density of 2.26 g/cm^3 and a water absorption rate of 6% leads to a 24% loss in CS (Kim et al. 2016).

The split tensile strength (STS) of concrete reduces with incorporation of coarse RCA as reported by the following studies. The STS value of RAC drops marginally by up to 9% (Kou et al. 2011; Gonzalez-Fonteboa et al. 2011). A higher reduction of about 14% (Yang et al. 2008) and 23% (Rao et al. 2011) was observed in the STS values of concrete made of coarse RCA. According to Sagoe-Crentsil et al. (2001), the STS of concrete is regulated by the binding mortar and not by the aggregate types. Improvement in the interface between old mortar layer of coarse RCA and fresh mortar leads to an increase in the STS of concrete. The interfaces present in RAC can be modified through using either the two-stage mixing (Tam et al. 2005) method or both primary and secondary crushing stages. The secondary crushing followed by the primary stage produces relatively rounded RCA with less adhered mortar (Pedro et al. 2017). Higher the replacement percentage of coarse RCA, lower is the STS of concrete (Fonseca et al. 2011; Gonzalez-Crominas and Etxeberria 2014). This can be related to an increase in the residual mortar content due to higher percentage replacement of coarse RCA in concrete. This, in turn, leads to a higher volume percentage of weak ITZ (Ibrahim et al. 2023).

The flexural tensile strength (FTS) of RAC was obtained within a variation of ± 15% as compared to that of NAC (Ravindrarajah and Tam 1985). Other research studies demonstrated a reduction of about 20% (Malhotra 1978), 24% (Bairagi et al. 1993), 5–20% based on different sources (Rao et al. 2017) and 7.5–13.8% (Yang et al. 2011). The ACI standard relationship between STS and FTS with CS, when applied to 100% RAC, was found to be unconservative (Tavakoli and Soroushian 1996). Silva et al. (2015) suggested similar relation of STS to CS in both NAC and RAC.

6.2 Effect of Coarse RCA on Elastic Modulus of Concrete

The factors, such as property of cement paste, aggregates, and ITZ, play major role in regulating the modulus of elasticity (E) of concrete. Specifically in the case of RAC, key parameters include aggregate replacement percentage, size, quality, mixing, mix design method, curing age, age of concrete, chemical admixture, and other additions (Silva et al. 2016). Coarse RCA has lower rigidity due to the attached mortar on its surface and greater tendency of deformability than natural aggregates. Higher coarse RCA replacement level implies more residual mortar content and lower elastic modulus. However, the quality of the mortar that is attached to the RCA also depends on the strength of the original aggregate and the procedure used for recycling. The E value of RAC is negatively affected by the complete substitution of coarse RCA. The elastic modulus of RAC reduces by 15% (Etxeberria et al. 2007; Rao et al. 2011), 18% (Gonzalez-Fonteboa et al. 2011), and 45% (Xiao et al. 2005) as compared to the modulus of NAC. The lower modulus of RAC can be attributed to mainly three factors such as weaker ITZ, higher porosity, and lesser modulus of coarse RCA itself. The reduction (17.7 and 22%) in the E value of RAC with respect to NAC is higher as compared to the decrease (12 and 15%) of compressive strength (Safiuddin et al. 2011 and Chen et al. 2003). According to Wang et al. (2021), the replacement of 100% coarse RCA results in a 40% loss in compressive strength, split tensile strength, and elastic modulus, and the flexural strength decreases by 20%.

6.3 Effect of Fly Ash on Compressive Strength of Coarse RCA-Based Concrete

In addition to the variables mentioned above, utilization of SCMs as partial cement substitute might potentially have a significant impact on the compressive strength of concrete made with coarse RCA. Besides the parameters such as replacement percentage of SCMs and coarse RCA, role of other factors that have been extensively studied for their impact on the compressive strength of concrete includes curing age, water-to-binder ratio, addition of superplasticizer, and mix design method. Kou et al.

(2007) examined the effect of using 25 and 35% fly ash in RAC with varying amount of coarse RCA (20, 50, and 100%). They observed a slight reduction (6%–9%) in 28-day CS but CS of 90 days cured RAC increased due to the presence of fly ash. On the other hand, Ann et al. (2008) reported that the CS of concrete reduces up to 90 days when 30% of the fly ash was used as a replacement for cement. However, higher CS is observed after 180 days curing. Somna et al. (2012) examined the combined effects of water-to-binder (w/b) ratio and fly ash replacement in RAC. When 35% of the cement was replaced with fly ash and the w/b ratio was 0.45, or when 20% fly ash was used with a higher w/b of 0.55 or 0.65, the compressive strength of the resulting RAC was either the same or higher compared to RAC without fly ash. If higher w/b ratio of 0.65 along with higher fly ash replacement of 50% is used in RAC, it hinders the pozzolanic reaction of fly ash owing to the predominant dilution effect. Addition of superplasticizer in 20% fly ash incorporated RAC can be useful in choosing a suitable w/c which balances the compressive strength and workability of the concrete (Saravana kumar and Dhinakaran 2013). The inclusion of 40% fly ash in RAC led to increased workability as well as compressive strength at both 56 and 91 days (Ahmed 2013). Utilizing high volumes of fly ash as a replacement of 40, 50, and 60% in RAC resulted in a decrease in CS. The combination of 50% fly ash and 50% RCA compromised 50% strength, but it showed a significant cost savings of 40% (Saravana kumar and Dhinakaran 2013).

Sunayana and Barai (2017) performed an extensive study on the impact of low-volume fly ash, specifically up to 30%, on the compressive strength of RAC. The concrete comprising 100% coarse RCA and up to 30% fly ash is denoted as FARAC in this book. They examined various parameters, including PPM mix design, percentage of fly ash replacement, w/b, and curing age. The w/b ratio was varied as 0.4, 0.43, 0.45, and 0.47. Additionally, the curing durations of 7, 28, and 90 days were chosen to examine the progression of strength at both early and later stages. The influence of w/b ratio on the CS of NAC, RFA20 (RCA with 20% fly ash), and RFA30 (RCA with 30% fly ash) after 28 days curing is presented in Fig. 6.1. Higher w/b ratios result in a clear reduction in CS due to a rise in porosity, regardless of the curing ages. However, it is worth noting that the rate of decrease of CS is less pronounced in FARAC as compared to NAC. At w/b ratio of 0.4, CS of RFA20 and RFA30 concrete mixes reduced by around 5.3 and 9.7% as compared to that of NAC. As the w/b ratio increased to 0.43 and 0.45, the effect of FARAC on the 28 days CS became less significant. The particle packing method (PPM) of mix design ensured a fixed proportion of aggregate for a particular packing density. Consequently, at higher w/b ratio, the amount of cement in the mix decreases which in turn reduces the CS of concrete. Nevertheless, the rate of CS reduction was lower in FARAC, due to the pozzolanic reaction in the presence of fly ash.

Let us analyse the effect of w/b ratio and curing ages on the compressive strength of NAC, RFA20, and RFA30 concrete mixes as depicted in Figs. 6.2, 6.3, 6.4 and 6.5. Figure 6.2 shows that the 7-day CS of NAC varied between 0.58 and 0.65 times the 28-day strength. A CS increase of about 27–33% was obtained in the 90 days cured concrete. The rate of reduction of CS with respect to w/b ratio was not very pronounced at early hydration ages. But, at 28 and 90 days, the strength experienced a

Fig. 6.1 Effect of w/b ratio on 28 days compressive strength

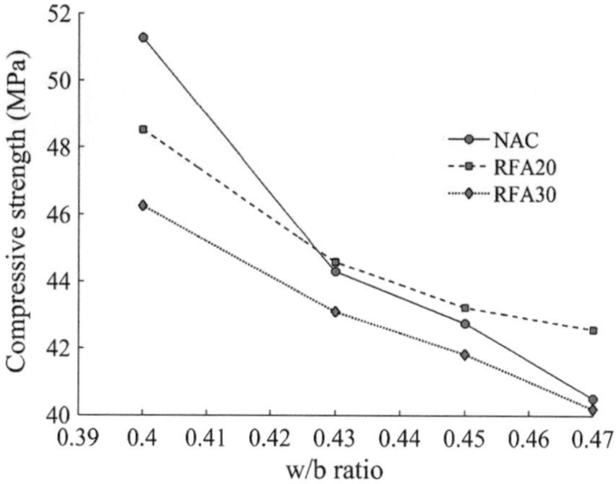

decline of 13% and 10%, respectively; this decrease in strength occurred when the w/b ratio increased from 0.4 to 0.43. It should be noted that this effect was not significant beyond 0.43. The CS of RFA20, as represented in Fig. 6.3, indicates that the 7-day strength was 0.52–0.55 times the 28-day strength, whereas the development of strength from 28 to 90 days was 16–20% of the 28-day strength. The rate of reduction of CS with increasing w/b ratio from (0.4 to 0.43) was reported as 11%, 8%, and 10% for 7-day, 28-day, and 90-day, respectively. The influence of w/b ratio on CS of RFA30 is illustrated in Fig. 6.4. After 7-days curing, 42–50% compressive strength was developed with respect to the 28 days strength. In the 90 days cured concrete, there was an increase of 11–13% CS as compared to the 28-day strength. The slower rate of strength development at 7 days in RFA20 and RFA30 concrete suggests that hydration reactions occurring at an early age are slow due to the presence of fly ash. The concrete in this study was prepared to have a characteristic compressive strength of 30 MPa with a target mean strength of 38 MPa. As previously discussed in Chap. 3, the concrete mix designed with PPM contains lower paste content and makes a stiffer mix. Target workability and comparable CS were obtained for NAC and FARAC at the w/b ratio of 0.45. Hence, 0.45 w/b ratio was considered in the preparation of concrete for further investigation of material and structural components covered in the following chapters.

Figure 6.5 illustrates the 28-day CS of NAC and RAC, both with and without fly ash, at a w/b ratio of 0.45. After 28 days curing, the FARAC mixes showed CS which are mostly equivalent to that of CS of NAC. However, the 7-day CS of RFA20 and RFA30 concrete were measured as 0.89 and 0.69 times the CS of NAC. The 90-day CS was found to be 0.90 and 0.83 times the corresponding CS of NAC. The reduction in the early strength of FARAC was ascribed to slower hydration reaction of fly ash. The hydration process of fly ash commences after the water reaches a pH level of 13, indicating the presence of enough alkalinity. Alkalinity is caused by the hydration of OPC, as stated by Neville (1995). The impact of using 100% coarse RCA in concrete was evaluated by comparing the performance of NFA20 with RFA20 concrete, as

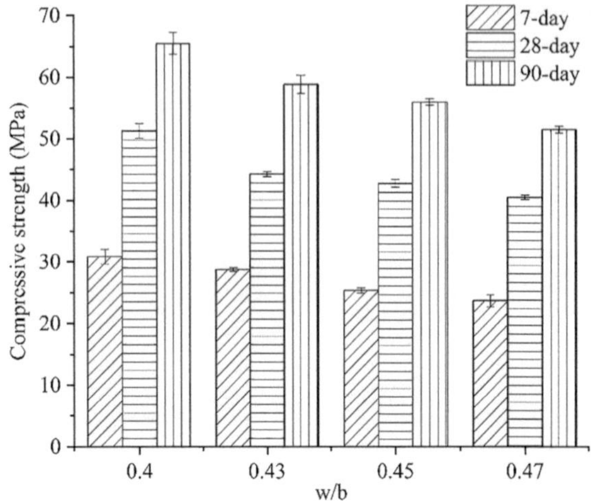

Fig. 6.2 Compressive strength of NAC at different curing ages and w/b

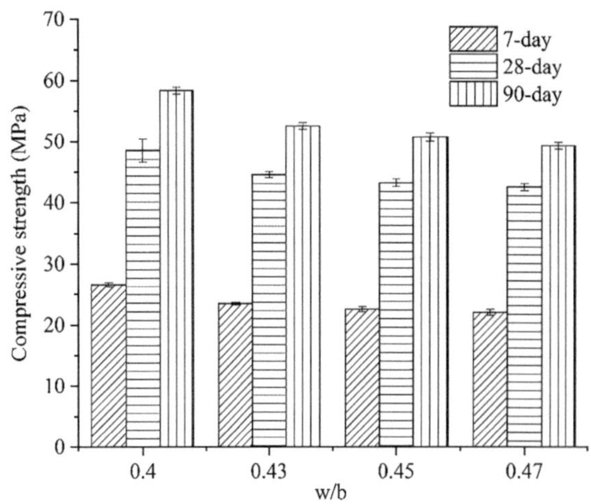

Fig. 6.3 Compressive strength of RFA20 at different curing ages and w/b

well as NFA30 with RFA30 concrete. The impact of RCA was notable during the early age, where the 7-day strength of RFA20 and RFA30 was respectively 0.90 and 0.85 times the strength of corresponding NFA20 and NFA30 concrete. This may be related to the existence of adhered mortar content on recycled coarse aggregate. The increase in 90 days strength of FARAC, though lesser, but significant as compared to NAC with and without fly ash at 28 days.

The variabilities of FARAC are inherent due to the source of RCA and nature of coal used in the operation of power station. The compressive strength of 30 samples of RFA20 and RFA30 is represented in terms of the frequency analysis histogram. A normal distribution curve is plotted (Figs. 6.6 and 6.7). Mean, standard deviation (SD), and coefficient of variation (COV) of RFA20 concrete were observed as 42.86, 2.85, 6.6, respectively, while for RFA30 concrete, the corresponding values were

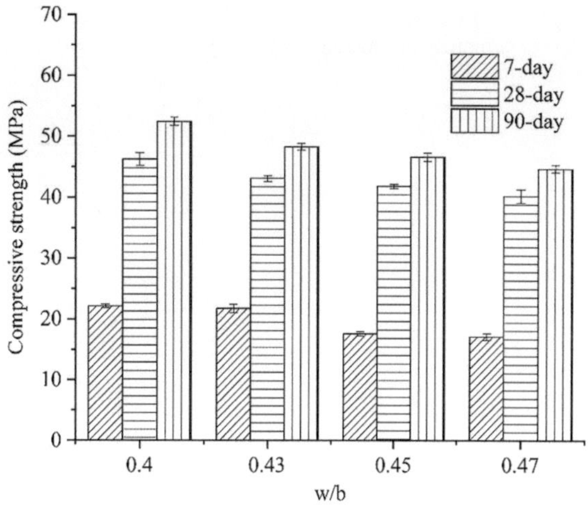

Fig. 6.4 Compressive strength of RFA30 at different curing ages and w/b

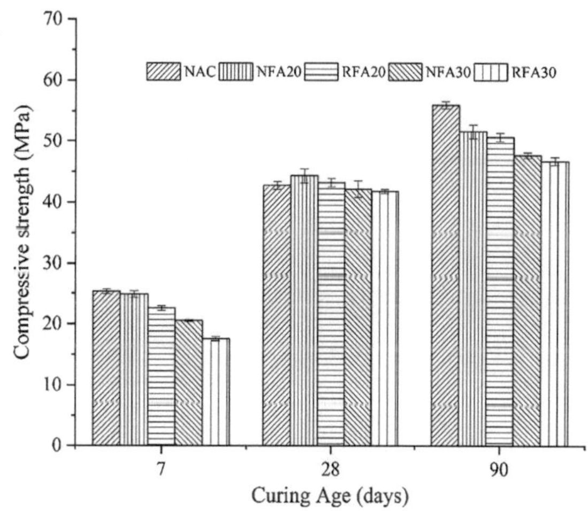

Fig. 6.5 Comparison of compressive strength at different curing ages

41.22, 1.9, 4.6. Higher COV indicated greater scattering of results implying uncertainty associated with RAC due to the quality of adhered mortar and microcracks in coarse RCA.

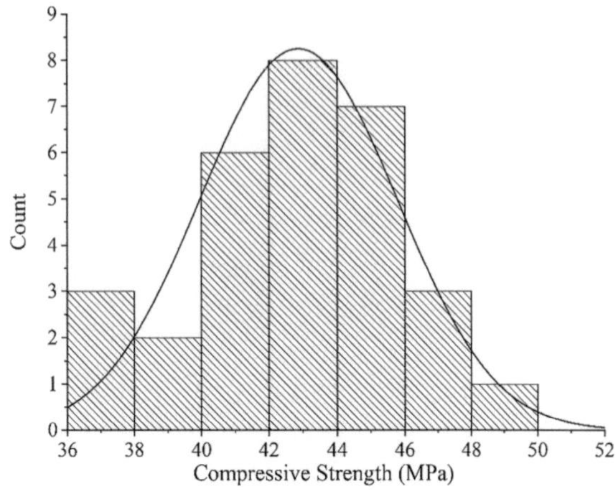

Fig. 6.6 Normal distribution plot for compressive strength of RFA20

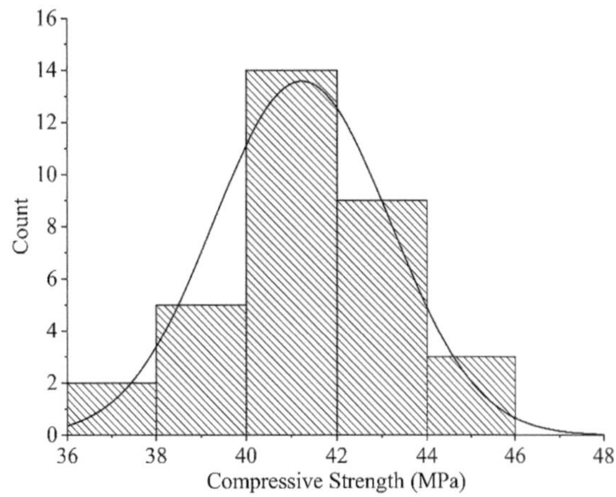

Fig. 6.7 Normal distribution plot for compressive strength of RFA30

6.4 Effect of Slag and Silica Fume on Compressive Strength of Coarse RCA-Based Concrete

The application of ground granulated blast furnace slag (GGBFS) in RAC has been shown to have positive effects on the compressive strength of concrete specimens cured for 28 days (Tüfekçi and Çakır 2017). They reported that 60% GGBFS can improve the compressive strength by 2.8, 7.4, and 8.9% as compared to NAC with 60% slag), in concrete mixes with 25, 50, and 100% RCA-based concrete. The highest strength gain was observed in concrete mixes with 100% coarse RCA. However, this effect was not observed in 7-days cured concrete. GGBFS was found to be beneficial in enhancing the properties of ITZ, that is the bond between the fresh paste and coarse RCA. This leads to filling of the microcracks in the adhered mortar layer in the long term. However, another study (Moghadam et al. 2021) reported an inverse relationship between GGBFS replacement (>30%) and compressive strength of RAC. Additionally, it was found that 10% silica fume contributed more towards the compressive strength development of RAC as compared to 20% fly ash or 30% GGBFS due to finer particles of silica fume (Qureshi et al. 2020).

6.5 Effect of Fly Ash on Tensile Strength of Coarse RCA-Based Concrete

Reduction in split tensile strength (STS) was observed when 25 and 35% of fly ash were used as cement replacement in RAC, as compared to RAC without fly ash and NAC (Kou et al. 2007). Nevertheless, the STS of RAC, both with and without fly ash, was found to be equal in 90 days cured concrete. Ann et al. (2008) found a similar pattern of decreased tensile strength in RAC that was integrated with fly ash. The ratio of STS to CS was comparable regardless of the type of binder or aggregates used. Ahmed (2013) also supported this observation of lower STS in RAC with and without fly ash. The measured STS was 6.5–11% of CS. According to Li et al. (2009), the flexural tensile strength (FTS) increased when fly ash and silica fume were present, perhaps because the paste had a higher packing density. Ahmed (2013) reported a reduction of FTS when 40% fly ash is used in RAC.

In their study, Sunayana and Barai (2017) examined the indirect tensile strength of NAC and FARAC in terms of STS and FTS. The results are depicted in Fig. 6.8 where mean and the standard deviation of 15 samples of each concrete type are also plotted. It is observed that STS of RFA20 and RFA30 concrete reduced by about 21 and 22% as compared to NAC. The adverse effect of fly ash replacement beyond 20% was not very significant in 100% RCA constituted concrete. FTS of RFA20 and RFA30 concrete was comparable to NAC with a variation less than 10%. STS of FARAC was 6% of the corresponding CS at 28 days. However, FTS was obtained as 11% of the respective compressive strength in FARAC. The rate of development of split tensile strength with compressive strength was slower as compared to the flexural strength in 28 days cured FARAC. Ahmed (2013) observed lower STS to CS ratio in 40% fly ash incorporated RAC.

The difference between FTS and STS of a particular concrete is due to the difference in the concrete volume under tension (Bentz 2019) which gives rise to the probabilistic size effect in concrete. As smaller concrete volume is under tension in the flexure test, the effect of modified aggregate-mortar interface (due to triple mixing method) becomes significant in FARAC. Thus, FTS was less affected by incorporation of fly ash and 100% RCA as compared to its STS. Application of 100% coarse RCA attributed to the reduction in STS and fly ash did not affect it further.

The ratio of STS to CS was evaluated by performing regression analysis on the available experimental data for 100% RCA replaced concrete (Butler et al. 2013; Guerra et al. 2014; Bairagi et al. 1993; Kim and Yun 2013; Yang et al. 2008; Thomas et al. 2013; Tabsh and Abdelfatah 2009; Gómez-Soberón 2002; Reis et al. 2015;

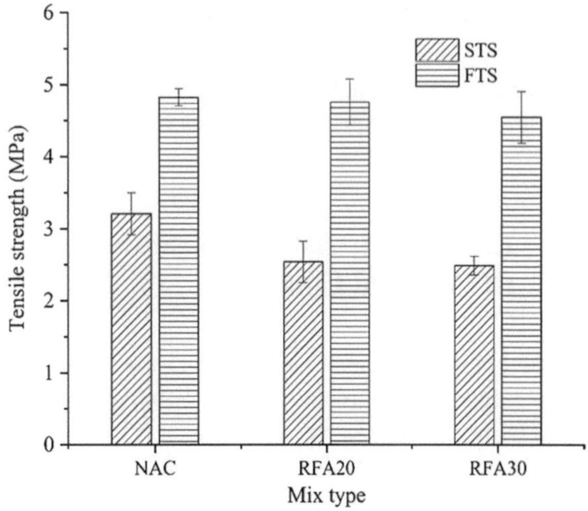

Fig. 6.8 Split and flexural tensile strength based on concrete types

Breccolotti and Materazzi 2013; Fonseca et al. 2011; Knaack and Kurama 2014; Sarhat and Sherwood, 2012; Folino and Xargay 2014). Data points corresponding to 100% RCA with fly ash include present experimental results along with the data from previous literature studies (Kou et al. 2007; Ann et al. 2008; Tangchirapat et al. 2010; Ahmed 2013; Saravana Kumar and Dhinakaran 2013; Çakır 2014). Figure 6.9 represents the STS/CS ratio for RAC with and without fly ash along with the standard relation for NAC proposed by Anoglu et al. (2006). Goodness of fit was verified through determination of integral absolute error (IAE). It represents deviation of the measured quantity from that of predicted. Power function regression equation was proposed for RAC. IAE was calculated using Eq. 6.1. The database included variations in cement type, quantity, mix design, recycled coarse aggregate properties, w/b ratio, fly ash quality as well as percentage replacement and use of water-reducing admixture. For the database of RAC with and without fly ash, IAE value was obtained as 10.41% and the correlation coefficient as 0.8.

$$\mathrm{IAE} = \sum_{i=1}^{n} \frac{(\mathrm{measured}_i - \mathrm{predicted}_i)^{0.5}}{\sum \mathrm{measured}_i} \tag{6.1}$$

The greater variability in the characteristics of RCA was evident in the increased dispersion of the data. The plot of the STS/CS ratio showed a declining trend as the cylinder compressive strength increased for both NAC and RAC. The proposed equation showed an enhanced tensile strength of FARAC as compared to NAC at lower CS (up to 25 MPa). However, the split tensile strength of RAC showed a pronounced decline at a higher rate for concrete with compressive strength beyond 25 MPa. Such pattern of the regression fit curve of RAC indicated poor tensile behaviour for high strength recycled aggregate concrete. The equation proposed previously for NAC yielded a wide variation in the results when applied to the RAC database. Thus,

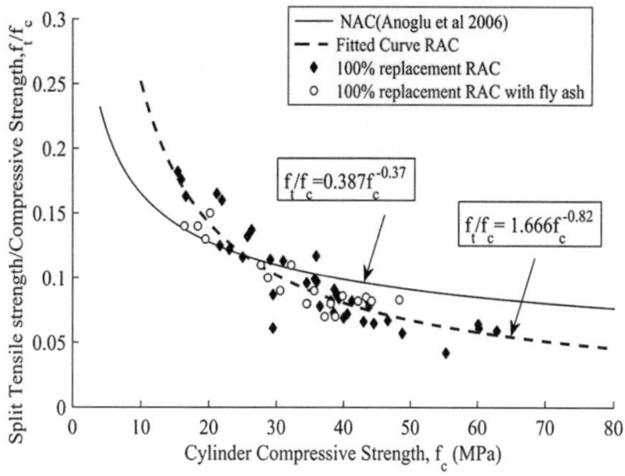

Fig. 6.9 Ratio of split tensile to compressive strength versus cylinder compressive strength

the present statistical analysis showed a realistic behaviour for STS to CS ratio of RAC with and without fly ash.

6.6 Effect of Slag and Silica Fume on Tensile Strength of Coarse RCA-Based Concrete

Application of GGBFS as a cement substitute at 30 and 60% in coarse RCA-based concrete showed positive effects on the tensile strength as compared to natural aggregate concrete (Tüfekçi and Çakır 2017). They also reported higher STS to CS ratio (0.072, 0.081, and 0.083) in 30% GGBFS substituted RAC as compared to the ratio (0.071, 0.074, and 0.075) in RAC without GGBFS (RCA replacement used were 25, 50, and 100%). Similar effects of higher STS to CS ratio were also observed in 60% GGBFS. The effect of silica fume, GGBFS and fly ash on split tensile strength of RAC was also studied by Qureshi et al. 2020. They observed an increase of STS by 8–14%, 2–10%, and 1–7% respectively in silicafume, GGBFS and fly ash incorporated 90 days cured RAC. This effect was attributed to strengthening of binder matrix and ITZ due to finer particle size and pozzolanic reactions of SCMs. On the other hand, another study by Mogadham et al. (2021) reported a reduction of (less than 15%) STS in 100% coarse RCA with 45% GGBFS concrete and about 22% decrease in flexural tensile strength.

6.7 Effect of Fly Ash on Modulus of Elasticity (E) of Coarse RCA-Based Concrete

Kou et al. (2007) observed reduction in E with increasing RCA replacement, but fly ash incorporation in RAC did not cause any further adverse effect on E. Modulus of elasticity of RAC with and without fly ash was lower than NAC by 16–20% due to the lower modulus of RCA (Somna et al. 2012), even though the compressive strength was equivalent. Modulus of elasticity of concrete as suggested by Thangchirapat et al. (2008) depends more on the aggregates rather than properties of cement paste; thus, having lesser effect of mineral addition on E. Sunayana and Barai (2017) investigated the effect of upto 30% fly ash and 100% coarse RCA on modulus of concrete where the mix design was carried out with particle packing method. A total 18 numbers of cylinder specimens were tested under compression to obtain the stress–strain relations. E values expressed in terms of $E/f_{cu}^{0.5}$ (f_{cu} represents cube compressive strength) are shown in Table 6.1. The ratios were found in the range of 5124–5709, 5357–5508, and 4461–5114 respectively for NAC, RFA20, and RFA30. The mean E values were 36 GPa, 35.16 GPa, and 31.34 GPa for NAC, RFA20, and RFA30, respectively. 5–10% reduction in modulus of elasticity was observed in FARAC as compared to NAC. This contrasts with higher reduction of E (15–30%) reported by Somna et al. (2012). It clearly demonstrated improvement in E due to PPM and triple mixing method. At 100% RCA, fly ash did not cause significant adverse impact on modulus of elasticity of concrete. This confirmed the observation already made by Kou et al. (2007) in this regard. Slight reduction in E values of RFA20 and RFA30 was attributed to lower modulus of RCA due to the presence of adhered mortar content.

6.8 Effect of Slag and Silica Fume on Elastic Modulus of Coarse RCA-Based Concrete

Silica fume used at 10% cement substitute increases (9–15%) modulus of elasticity of RAC due to refinement of pore size and extent of pozzolanic reaction. On the other hand, fly ash and GGBFS need longer curing time (90 days) for about 9% increase in elastic modulus of RAC (Qureshi et al. 2020).

6.9 Summary

In this chapter, the hardened properties of coarse RCA and cementitious materials incorporated into concrete are discussed. More specifically, use of low-volume fly ash in 100% coarse RCA based concrete is evaluated. The key findings can be summarized as follows:

Table 6.1 Modulus of elasticity of concrete

Concrete mix	Compressive strength (MPa)	E (MPa)	$E/f_{cu}^{0.5}$
NAC	42.75	33,500	5124
	45.11	34,700	5166
	47.50	35,400	5136
	48.24	36,400	5241
	42.65	33,700	5160
	45.07	37,700	5616
	43.68	36,900	5583
	44.77	38,200	5709
RFA20	36.76	33,390	5507
	40.18	34,130	5384
	42	34,720	5357
	42.46	35,590	5462
	43.22	36,210	5508
RFA30	40.3	28,320	4461
	40.49	29,400	4620
	42.7	31,410	4807
	43.82	31,910	4819
	44.31	35,660	5114

- Long-term compressive strength increased significantly for FARAC, though the increase was lesser than that in NAC. Flexural strength is not much affected by the incorporation of fly ash in RAC.
- Comparable modulus of elasticity of NAC and FARAC indicates an improvement due to the particle packing method of mix design.
- Split tensile strength of FARAC reduced significantly, and the ratio of STS to CS decreased beyond 25 MPa as compared to NAC.
- Silica fume at 10% substitution and GGBFS (30 or 60%) may be beneficial in compressive strength, and tensile strength loss of RAC. However, longer curing period of 90 days is recommended for achieving this positive effect of cementitious materials.
- Fly ash and GGBFS can improve workability of RAC provided the particles are spherical which improve lubrication of cementitious paste.

References

Ahmed SFU (2013) Properties of concrete containing construction demolition wastes and fly ash. J Mater Civ Eng 25(12):1864–1870
Ann K, Moon H, Kim Y, Ryou J (2008) Durability of recycled aggregate concrete using pozzolanic materials. Waste Manage 28(6):993–999
Anoglu N, Girgin ZC, Arıoglu E (2006) Evaluation of ratio between splitting tensile strength and compressive strength for concretes up to 120 MPa and its application in strength criterion. ACI Mater J 103(1):18–24
Bairagi NK, Ravande K, Pareek VK (1993) Behaviour of concrete with different proportions of natural and recycled aggregates. Resour Conserv Recycl 9(1–2):109–126
Bentz EC (2019) Empirical modeling of cracking in reinforced concrete. ACI Struct J 116(3):233–242
Breccolotti M, Materazzi AL (2013) Structural reliability of bonding between steel rebars and recycled aggregate concrete. Constr Build Mater 47:927–934
Butler L, West JS, Tighe SL (2013) Effect of recycled concrete coarse aggregate from multiple sources on the hardened properties of concrete with equivalent compressive strength. Constr Build Mater 47:1292–1301
Çakır Ö (2014) Experimental analysis of properties of recycled coarse aggregate (RCA) concrete with mineral additives. Constr Build Mater 68:17–25
Chen HJ, Yen T, Chen KH (2003) Use of building rubbles as recycled aggregates. Cem Concr Res 33(1):125–132
Etxeberria M, Vázquez E, Marí A, Barra M (2007) Influence of amount of recycled coarse aggregates and production process on properties of recycled aggregate concrete. Cem Concr Res 37(5):735–742
Folino P, Xargay H (2014) Recycled aggregate concrete–mechanical behavior under uniaxial and triaxial compression. Constr Build Mater 56:21–31
Fonseca N, De Brito J, Evangelista L (2011) The influence of curing conditions on the mechanical performance of concrete made with recycled concrete waste. Cement Concr Compos 33(6):637–643
Giergiczny Z (2019) Fly ash and slag. Cem Concr Res 124:105826
Gómez-Soberón JM (2002) Porosity of recycled concrete with substitution of recycled concrete aggregate: an experimental study. Cem Concr Res 32(8):1301–1311
Gonzalez-Crominas A, Etxeberria M (2014) Experimental analysis of properties of high performance recycled aggregate concrete. Constr Build Mater 52:227–235
Gonzalez-Fonteboa B, Martinez-Abella F (2005) Recycled aggregates concrete: aggregates and mix properties. Materiales De Construction 55(279):53–66
Gonzalez-Fonteboa B, Martinez-Abella F, Eiras-Lopez J, Seara-Paz S (2011) Effect of recycled coarse aggregate on damage of recycled concrete. Mater Struct 44(10):1759–1770
Guerra M, Ceia F, De Brito J, Júlio E (2014) Anchorage of steel rebars to recycled aggregates concrete. Constr Build Mater 72:113–123
Kim SW, Yun HD (2013) Influence of recycled coarse aggregates on the bond behavior of deformed bars in concrete. Eng Struct 48:133–143
Kim N, Kim J, Yang S (2016) Mechanical strength properties of RCA concrete made by a modified EMV method. Sustainability 8:1–15
Knaack AM, Kurama YC (2014) Behavior of reinforced concrete beams with recycled concrete coarse aggregates. J Struct Eng 141(3):B4014009
Kou SC, Poon CS, Chan D (2007) Influence of fly ash as cement replacement on the properties of recycled aggregate concrete. J Mater Civ Eng 19(9):709–717
Kou SC, Poon CS, Agrela F (2011) Comparisons of natural and recycled aggregate concretes prepared with the addition of different mineral admixtures. Cement Concr Compos 33(8):788–795

Ibrahim M, Alimi W, Assaggaf R, Salami BA, Oladapo EA (2023) An overview of factors influencing the properties of concrete incorporating construction and demolition wastes. Constr Build Mater 367:130307

Li J, Xiao H, Zhou Y (2009) Influence of coating recycled aggregate surface with pozzolanic powder on properties of recycled aggregate concrete. Constr Build Mater 23(3):1287–1291

Malhotra VM (1978) Use of recycled concrete as new aggregate. In: Symposium on energy and resource conservation in the concrete industry, CANMET report no. 76–8, CANMET, Ottawa, Canada

Moghadam AS, Omidinasab F, Goodarzi SM (2021) Characterization of concrete containing RCA and GGBFS: mechanical, microstructural, and environmental properties. Constr Build Mater 289:123134

Neville AM (1995) Properties of concrete, vol 4. Longman, London

Pedro D, de Brito J, Evangelista L (2017) Evaluation of high-performance concrete with recycled aggregates: use of densified silica fume as cement replacement. Constr Build Mater 147(2017):803–814

Qureshi LA, Ali B, Ali A (2020) Combined effects of supplementary cementitious materials (silica fume, GGBS, fly ash and rice husk ash) and steel fiber on the hardened properties of recycled aggregate concrete. Constr Build Mater 263:120636

Rahal K (2007) Mechanical properties of concrete with recycled coarse aggregate. Build Environ 42(1):407–415

Rao MC, Bhattacharyya SK, Barai SV (2011) Behaviour of recycled aggregate concrete under drop weight impact load. Constr Build Mater 25(1):69–80

Rao MC, Bhattacharyya SK, Barai SV (2017) Influence of different demolished old structures concrete as aggregate on properties of concrete. Indian Concr J 91:68–85

Ravindrarajah R, Tam C (1985) Properties of concrete made with crushed concrete as coarse aggregate. Mag Concr Res 37(130):29–38

Reis N, de Brito J, Correia JR, Arruda MR (2015) Punching behaviour of concrete slabs incorporating coarse recycled concrete aggregates. Eng Struct 100:238–248

Safiuddin M, Alengaram UJ, Salam MA, Jumaat MZ, Jaafar FF, Saad HB (2011) Properties of high-workability concrete with recycled concrete aggregate. Mater Res 14(2):248–255

Sagoe-Crentsil KK, Brown T, Taylor AH (2001) Performance of concrete made with commercially produced coarse recycled concrete aggregate. Cem Concr Res 31(5):707–712

Saravana Kumar P, Dhinakaran G (2013) Strength characteristics of high-volume flyash–based recycled aggregate concrete. J Mater Civil Eng 1127–1133

Sarhat SR, Sherwood EG (2012) Residual mechanical response of recycled aggregate concrete after exposure to elevated temperatures. J Mater Civ Eng 25(11):1721–1730

Silva RV, de Brito JMCL, Dhir RK (2015) The influence of the use of recycled aggregates on the compressive strength of concrete: a review. Eur J Environ Civ Eng 19(7):825–849

Silva RV, De Brito J, Dhir RK (2016) Establishing a relationship between modulus of elasticity and compressive strength of recycled aggregate concrete. J Clean Prod 112:2171–2186

Somna R, Jaturapitakkul C, Chalee W, Rattanachu P (2012) Effect of the water to binder ratio and ground fly ash on properties of recycled aggregate concrete. J Mater Civ Eng 24(1):16–22

Sunayana S, Barai SV (2017) Recycled aggregate concrete incorporating fly ash: comparative study on particle packing and conventional method. Constr Build Mater 156:376–386

Tabsh SW, Abdelfatah AS (2009) Influence of recycled concrete aggregates on strength properties of concrete. Constr Build Mater 23(2):1163–1167

Tam VW, Gao X, Tam C (2005) Microstructural analysis of recycled aggregate concrete produced from two-stage mixing approach. Cem Concr Res 35(6):1195–1203

Tangchirapat W, Buranasing R, Jaturapitakkul C, Chindaprasirt P (2008) Influence of rice husk–bark ash on mechanical properties of concrete containing high amount of recycled aggregates. Constr Build Mater 22(8):1812–1819

Tangchirapat W, Buranasing R, Jaturapitakkul C (2010) Use of high fineness of fly ash to improve properties of recycled aggregate concrete. J Mater Civ Eng 22(6):565–571

Tavakoli M, Soroushian P (1996) Drying shrinkage behavior of recycled aggregate concrete. Concr Int 18(11):58–61

Thomas C, Setién J, Polanco JA, Alaejos P, De Juan MS (2013) Durability of recycled aggregate concrete. Constr Build Mater 40:1054–1065

Tüfekçi MM, Çakır Ö (2017) An investigation on mechanical and physical properties of recycled coarse aggregate (RCA) concrete with GGBFS. Int J Civil Eng 15:549–563

Wang B, Yan L, Fu Q, Kasal B (2021) A comprehensive review on recycled aggregate and recycled aggregate concrete. Resour Conserv Recycl 171:105565

Xiao J, Li J, Zhang C (2005) Mechanical properties of recycled aggregate concrete under uniaxial loading. Cem Concr Res 35(6):1187–1194

Yang KH, Chung HS, Ashour AF (2008) Influence of type and replacement level of recycled aggregates on concrete properties. ACI Mater J 105(3):289–296

Yang J, Du Q, Bao Y (2011) Concrete with recycled concrete aggregate and crushed clay bricks. Constr Build Mater 25(4):1935–1945

Chapter 7
Microstructure Properties

Preview

Concrete comprises of three different phases such as (1) aggregate, (2) hardened cement matrix and (3) interfacial transition zone (ITZ) between the aggregate and matrix. Heterogeneity in concrete at the microstructural level is caused mainly due to the non-uniform distribution of water in the zone near an aggregate edge and that farther from the aggregate that is in the matrix region. Phenomena which occur closer to an aggregate surface in concrete are: (1) availability of more water which facilitates formation of larger size capillary voids after hydration (2) growth of preferentially oriented layer of crystalline hydrated products (mainly calcium hydroxide) and (3) development of microcracks even prior to loading. Moreover, the microstructure rather than being fundamentally constant, varies with curing ages. Presence of aggregates in general (due to wall effect) and types of aggregate such as recycled coarse aggregate in particular result in a complicated and inhomogeneous microstructure than the plain cement paste. The layer of adhered mortar on the surface of recycled coarse aggregate makes the interface more complex.

In this chapter, we will try to find out answer to the following questions: How does incorporation of supplementary cementitious materials alter the aggregate-matrix interface in recycled aggregate concrete? How do the microstructural properties affect the macro behaviour of concrete?

7.1 Interfacial Transition Zone of RAC-Based Concrete

Figure 7.1 shows a comparison of concrete slices made of (a) natural coarse aggregates and (b) 100% recycled coarse aggregate. The recycled aggregate based concrete clearly shows the presence of two ITZs, i.e., one between parent aggregate and old mortar and another between the old and fresh mortar. The old ITZ is weaker due

S. Sunayana and S. V. Barai, *Green Concrete*, Springer Transactions in Civil and Environmental Engineering, https://doi.org/10.1007/978-981-96-6982-0_7

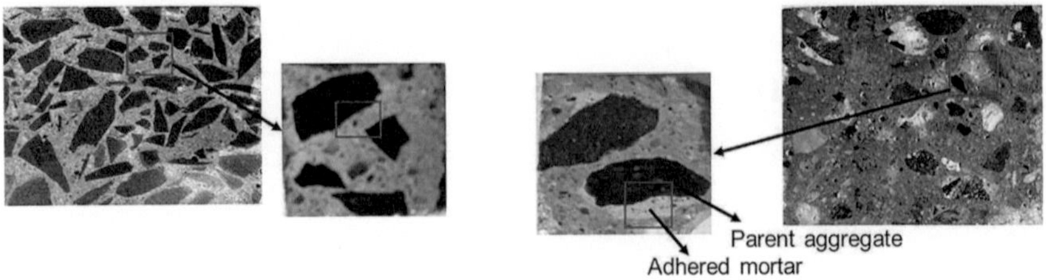

Fig. 7.1 Polished concrete slices showing layer of adhered mortar in RAC

to higher porosity in 50% RCA replaced concrete. Lower water to cement ratio or higher cement content can densify the ITZ in RAC (Etxeberria et al. 2006).

The microstructure of concrete is constituted of the hydration products, pores and unhydrated binder particles. The degree of hydration and distribution of hydration products regulates the micro-level properties. Some of the key experimental techniques used to assess these aspects of hydration include differential scanning calorimetry (DSC) or thermogravimetric (TG) or differential thermal analysis (DTA) and the scanning electron microscope (SEM) imaging. Baofan (2013) reported loose and porous microstructure of RAC through SEM imaging due to the presence of larger $Ca(OH)_2$ and ettringite crystals. Tam et al. (2009) conducted a DSC study and reported lower summation of heat values in RAC as compared to NAC. The microcracks present in the old ITZ absorbs some of the mixing water which lowers the water content available for hydration reaction. So, lesser hydration products are formed resulting higher porosity and lower strength of RAC. The rate of development of temperature within a concrete sample in a semi-adiabatic curing condition gives the measure of degree of hydration of paste in RAC. A linear relationship between the degree of hydration and the compressive strength of RAC given by Koenders et al. (2014) highlights the importance of moisture content of the aggregate.

The three-dimensional picture of the microstructure comprising spatial distribution of pores can be visualized using X-ray microtomography (XRT). Leite and Monteiro (2016) used this method and underlined the role of parent concrete strength and initial moisture conditions of RCA in regulating the properties of RAC. RCA coming from a lower strength parent concrete negatively affects the strength of RAC. RCA when used in dry state shows connected macropores in the XRT images unlike the aggregates used in surface saturated dry (SSD) condition. So, it is recommended to use the aggregates in SSD condition for lower porosity and pore connectivity.

Role of Mixing Methods in Modifying the Microstructure of RAC

The pre-mix procedure two-stage mixing approach (TSMA, discussed in Chap. 3) which develops some strength enhancing hydration products and hence beneficial in improving the quality of RAC. Li et al. (2012) observed the new ITZ under SEM and also reports the effectiveness of the TSMA in making the new ITZ denser and more homogenous. They also supported this argument by performing a nano-indentation study which shows higher modulus of ITZ as compared to the concrete prepared with normal mixing method. The triple mixing method (discussed in Chap. 3) for mixing of constituents is also helpful for making denser ITZ (Kong et al. 2010). The fly ash was used as a 5% cement replacement in this study. More C–S–H is formed in the interface of fly ash coated RCA due to the pozzolanic reaction between fly ash and larger CH particles.

With this background, we will now discuss the role of SCMs on the microstructure of recycled aggregate based concrete. Firstly, we will focus on the role of fly ash in RAC. We will try to find answer to the following question: How do the microstructure of RAC comprising 100% recycled coarse aggregates alter if we use low-volume fly ash as partial cement substitute in the concrete?

The micro-level properties of fly ash incorporated RAC have been studied previously by Sunayana and Barai (2021) using the following experimental techniques.

- Investigation of degree of hydration reactions through thermo-gravimetric analysis. (TGA).
- Observation of local variations in the microstructure specifically focussing on the morphology of hydration products through qualitative and quantitative analysis of SEM images.
- Estimation of micro-mechanical properties ITZ properties such as hardness and modulus through nano-indentation test.
- Three-dimensional porosity estimation through X-ray microtomography test.

7.2 Sample Preparation and Working Principle

Differential thermal analysis (DTA) and Thermogravimetric analysis (TG) were performed by Sunayana and Barai (2021) to investigate the effects of fly ash on the degree of hydration of FARAC. Sample preparation and details of the instrument are discussed below.

7.2.1 Thermogravimetric Analysis (TGA)

Cube specimens were prepared for the purpose of conducting the compressive strength tests. Once the specimens had undergone the process of curing and hardening, the cement paste that had adhered to the surface of the aggregates was gathered. The powder obtained from FARAC had a mixture of old and fresh mortar, while NAC consisted solely the fresh paste created after hydration. The analysis focused on a 20 mg sample of the powder, which had particles smaller than 75 μm in size.

Pyris diamond thermal analyser instrument as shown in Fig. 7.2 was used in this study. The powder concrete sample undergoes heating from an initial temperature of 30 °C to a final temperature of 800 °C, with a heating rate of 10 °C per minute, in a nitrogen atmosphere. The specified working temperature range corresponds to room temperature up to that when there is negligible or no mass loss.

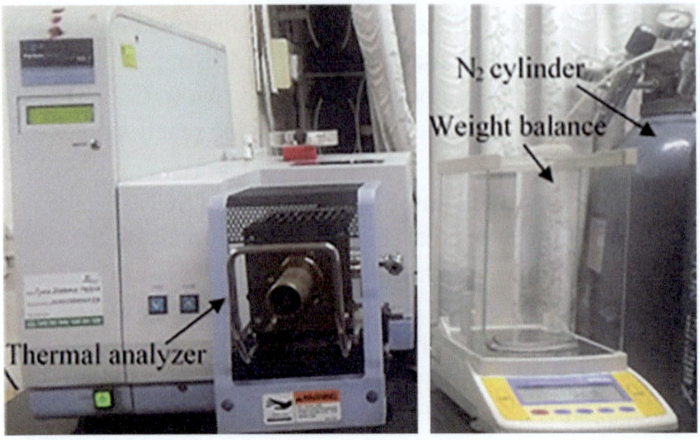

Fig. 7.2 Thermal analyzer

The measurement of mass loss is conducted using a TG balance with a precision of 0.1 mg. Aluminium (Al) was considered as the reference material because it is thermally inert. That means when Al is subjected to temperature rise, it undergoes only mass loss without any changes of material phases. The DTA is computed as the temperature difference between the concrete sample and the reference material. Upon heating, cement paste undergoes dissociation, resulting in the release of H_2O and CO_2 which were consumed during the hydration reaction. This process is indicated by the mass loss or a TG value. The first derivative of mass loss with respect to time is termed as derivative thermo-gravimetry (DTG) representing the rate of mass loss. Such process of using TG and DTG curve simultaneously to study the mass loss was suggested by Marsh and Day (1988). Area under a peak of the DTA curve was proportional to the enthalpy change of the reaction which occurred at that temperature range. According to the first law of thermodynamics, enthalpy is defined as the heat change of a system when it absorbs heat at a constant pressure.

Dehydration and dehydroxylation reactions were associated with the direct release of chemically combined water used up during the hydration. Decarbonation reaction released CO_2 by break down of calcite. These reactions can be identified based on the temperature range of the peak of DTA or DTG curve (Monteagudo et al. 2014).

(i) Dehydration of C–S–H and other hydrates at a temperature range of about 105–430 °C. Primary hydration (Vedalaxmi et al. 2003) of cement is expressed in the Eq. 7.1

$$2(3CaO \cdot SiO_2) + 6H_2O \xrightarrow{\text{yields}} 3CaO \cdot 2SiO_2 \cdot 3H_2O + 3Ca(OH)_2 \quad (7.1)$$

Secondary hydration reactions occur when the silica in fly ash reacts with the calcium hydroxide produced during primary hydration.

$$3Ca(OH)_2 + 2SiO_2 + H_2O \xrightarrow{\text{yields}} 3CaO \cdot 2SiO_2 \cdot 3H_2O \quad (7.2)$$

(ii) Dehydroxylation of CH: Dehydroxylation reaction as shown in Eq. 7.3 occurs in the temperature ranges around 430–530 °C.

$$Ca(OH)_2 \xrightarrow{\text{yields}} CaO + H_2O \tag{7.3}$$

(iii) Decarbonation: The CH molecule generated during hydration combines with ambient CO_2 in the presence of moisture, resulting in the formation of calcium carbonate or $CaCO_3$. It may form even if sufficient care is taken while preparing the samples prior to testing (Shafiq and Nuruddin 2010). Thus, decarbonation occurs at a higher temperature of around 580°–1000 °C.

$$CaCO_3 \xrightarrow{\text{yields}} CaO + CO_2 \tag{7.4}$$

In the study (Sunayana and Barai 2021), the mass loss (%) was determined in relation to the mass at 105 °C. The reason for selecting this reference temperature is because at 100 °C, physically adsorbed water in the powdered cement paste sample evaporated when it was exposed to temperature. Chemically bonded water begins to evaporate at temperatures higher than 105 °C (Monteagudo et al. 2014). The degree of hydration can be estimated from the loss of chemically bound water. Quantity of CH formed is estimated using the equivalent molar weight of Ca $(OH)_2$.

$$Ca(OH)_2 = \left(\frac{74}{18}\right) \text{Area}_{\text{dehydroxylation}} \tag{7.5}$$

The chemically bound water giving a measure of the degree of hydration can be determined by analysing the components of dissociation reactions from the thermogravimetric curves. For this, the method suggested by Bhatty et al. (1986) is used as shown below in Eqs. 7.6 and 7.7. The results obtained using this method is discussed in the Sect. 7.6.

$$w_b = ldh + ldx + 0.41 * ldc \tag{7.6}$$

$$\alpha = \frac{w_b}{0.24} * 100 \tag{7.7}$$

where w_b = bound water content per unit of binder

α = degree of hydration (%)

ldh = dehydration component; ldx = dehydroxylation; and ldc = decarbonation component.

7.2.2 Scanning Electron Microscope Imaging

Polished concrete samples were subjected to SEM imaging to evaluate the surface structure of hydration products. Through quantitative image analysis, the distribution and area percentages of various phases, such as pores, CH, unhydrated cement, or fly ash particles are determined.

Sample Preparation: Polishing

After the designated 7, 28, and 90-day curing ages, cube and prism specimens were taken out from the curing tank and left to dry. The outer edges of the specimens were then removed from these specimens using a low-speed, precision diamond cutter. This is required to prevent variations at the specimen's upper surface due to carbonation or bleeding. At first, slices of about 10 mm thick were cut from the concrete specimens. These slices were then divided into smaller pieces, each measuring 0.2 cm in thickness and 1–2 cm in width. The procedure for preparing the sample was adopted from Diamond (2001). These slices are then dried in the desiccators at low temperature. Low viscosity epoxy raisin is vacuum impregnated into the dried specimens and left to set. The voids are filled with epoxy raisin, which can enhance the contrast between different phases of hydration products. The variation in the mean atomic number of the matter causes contrast differences in the images (Scrivener 2004). Accordingly, the lowest mean atomic number correlates to pores (black), and the highest mean atomic number denotes unhydrated cementitious material (bright). Surfaces that are flat and polished are necessary to reduce the amount of topographic contrast in SEM images. Carborundum dust and grits with 400, 600, and 1000 μ mesh sizes were utilized for polishing for around 30 min. Because of the variations in hardness and surface roughness, the aggregate surface needs to be polished carefully to prevent any damages in the matrix phase. Diamond paste of 3, 1 and 0.25 μ were applied and kept for a duration of 30 min to 1 h. Diamond paste with smaller diameters required less time to cure. Once each polishing step was finished, the samples were ultrasonically cleaned in alcohol. Before samples were tested under a SEM, they were coated with gold.

Using a Zeiss Merlin SEM, pictures were taken at 10 kV at various resolutions: 1000×, 3000×, 5000×, and 10,000×. In this study, the imaging techniques such as secondary electron (SE) and back scatter electron (BSE) imaging both were employed. In the BSE approach, the interaction between the electron beam and matter caused the electrons to scatter elastically from a deeper depth.

Image Analysis

The BSE images were processed using the software Image J. Image analysis was carried out in a MATLAB environment by adopting the procedure discussed in Brand (2015). Images were first transformed to a greyscale with intensities ranging from 0 to 255. Noise was eliminated by applying Lee's (1983) sigma averaging filter. To distinguish between various component phases, including pores, CH, other hydration products, and unhydrated cementitious material, the grey scale intensity histogram

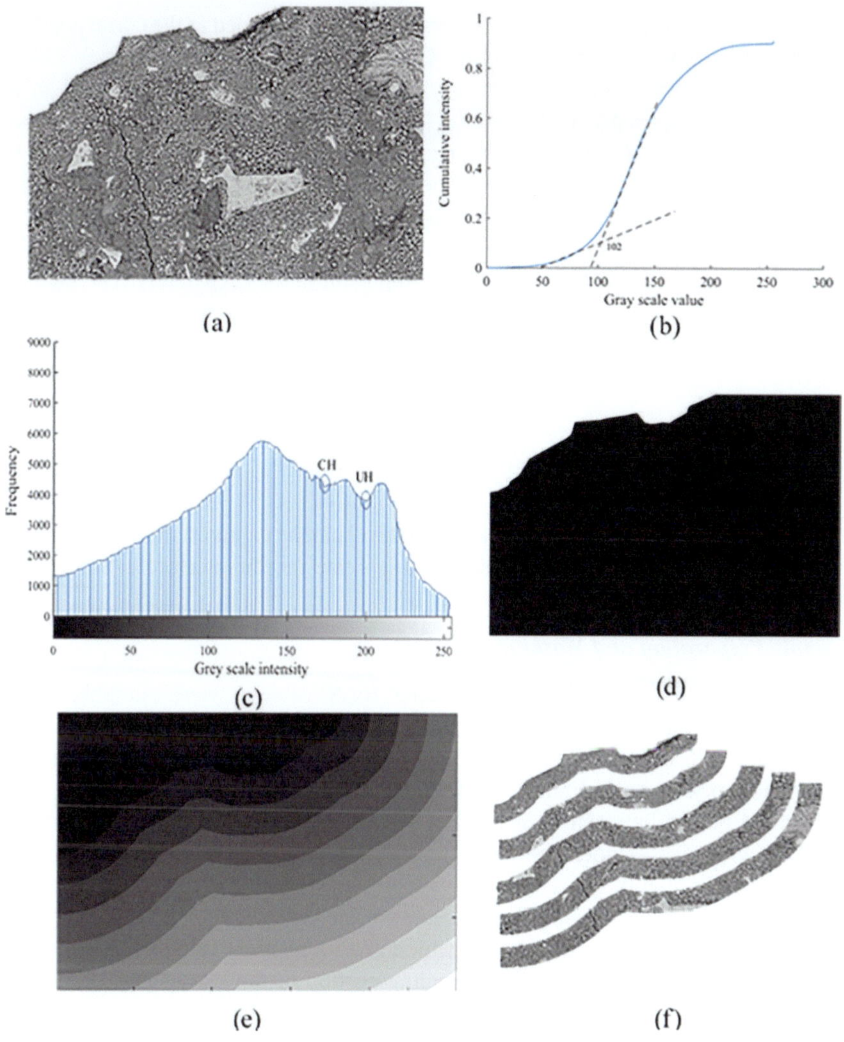

Fig. 7.3 Analysis of SEM images **a** original image after removing aggregate, **b** overflow point method (Wong et al. 2006) for thresholding of pores, **c** local minima identification for thresholding of CH and unhydrated products, **d** aggregate mask, **e** Euclidian distance mapping (Wong et al. 2006), **f** 10 μm strips

was plotted. To identify and determine the number of pixels corresponding to each component phase, segmentation and thresholding techniques were applied to the images. This procedure is illustrated in Fig. 7.3. Segmentation of pores was achieved through the overflow point method suggested by Wong et al. (2006). In the cumulative grey scale intensity plot, the intersection of two tangents corresponding to a change of slope was considered as the threshold grey scale intensity for pores. The local minima value in the pixel intensity histogram, which is regarded as the corresponding threshold point, was used to identify the CH and unhydrated components.

Since it was more difficult to clearly identify these peaks in concrete than in mortar samples, appropriate engineering judgement was used to ensure correctness. The appropriate threshold settings were applied in order to convert these images to binary. Isolated pores making up 1% of the image size were identified as noise. After the aggregate was removed, a binary mask was made for the aggregate component of the entire image. Next, the binary mask made for the aggregate was overlaid with the binary image displaying each of the component phases separately. The method of Euclidean distance mapping was utilized to measure 10 μm strips. Distribution of different phases along the aggregate-mortar interface were analysed in these strips.

7.2.3 Nanoindentation Study

Heterogeneity of concrete at microstructural level can be captured through measurement of micro- mechanical properties at the aggregate-mortar interface such as elastic modulus and hardness of the.

Sample Preparation and Working Principle

The polished sample same as that considered for SEM study were examined under the nano indenter. A three-sided pyramidal Berkovich probe with a tip radius of 100 nm and a tip angle of 142.3° was affixed to a hysitron triboindenter. A trapezoidal load function with a maximum load value of 1200 μN was applied for 10 s, with a 2 s hold period between each loading and unloading cycle. The average depth of penetration of the indenter was about 250 mm. At every indentation point, the load-penetration depth was continually monitored. Figure 7.4 represents a typical load-penetration curve. Average pressure below the indenter tip is determined using the applied load per unit contact area, which gives a measure of the hardness at the interface. In addition, slope of the unloading portion of the load-penetration curve denotes the reduced elastic modulus of the interface. Equation 7.8 can be used to compute modulus of concrete at the interface by substituting the Poisson's ratio value as 0.2.

$$\frac{1}{E_r} = \frac{1 - \nu^2}{E} + \frac{1 - \nu_{in}^2}{E_{in}} \tag{7.8}$$

where E and υ are material elastic constants and E_{in} and υ_{in} are indenter elastic constants. Specifications of the indenter used were as follows: $E_{in} = 1140$ GPa and $\upsilon_{in} = 0.07$. Curves having any discontinuity or with showing a high depth of penetration were discarded. The indentation was performed in the interfacial region with a size 10 × 3 grid with 15 μm spacing. Both the old interface (aggregate-adhered mortar) and new interface (adhered mortar and fresh mortar) were examined in FARAC. In general, the elastic modulus of individual phases such as porosity, C–S–H (along

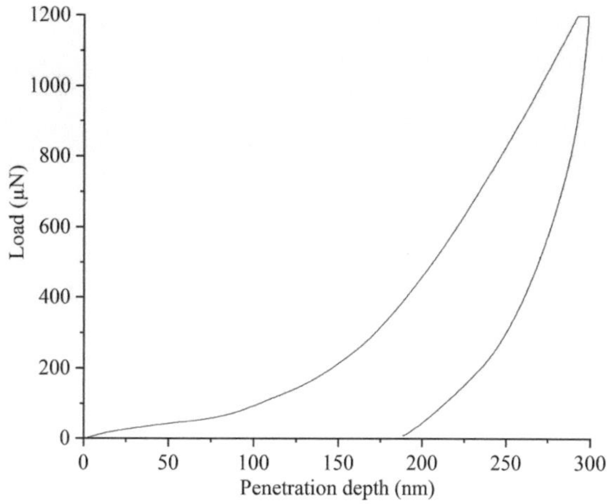

Fig. 7.4 Typical load–displacement (P–h) curve

with other hydration products), CH and unhydrated cement in a hydrated cement paste generally remain in the range of below 7 GPa, 7–34 GPa, 34–50 GPa and higher than 50 GPa respectively (Constantinides and Ulm 2004; Sorelli et al. 2008).

7.2.4 Non-destructive Scanning Using X-Ray Microtomography Test

Using non-destructive scanning techniques including X-ray computed microtomography, three-dimensional porosity, distribution, and pore shapes were assessed in NAC and FARAC samples.

Sample Preparation and Scanning Conditions

Concrete cylinder specimens of 200 mm height and 100 mm diameter were cast and put through testing for non-destructive scanning. In addition, to observe the distribution of pores of smaller sizes, cuboids of 1–2 cm were used for scanning. Furthermore, this non-destructive scanning requires no sample preparation, rendering it more advantageous than the microstructural tests that were previously discussed. Scanning of the cylinder specimens were performed at a resolution of voxel size 142 μm. The instrument used for this is displayed in Fig. 7.5. The specified parameters for voltage and current are considered as 150 kV and 100 μA respectively. A complete 3D image scan consisted of 1000 two-dimensional slices taken at 500 ms intervals. Using a 0.5 mm Cu filter, beam hardening corrections were applied to prevent noise or artifacts. By eliminating the specimen's edge effect, a specific area of the specimen was chosen for scanning based on the required field of view and resolution. The smaller cuboid specimens were scanned at a voxel size of 16 μm to 20 μm to capture

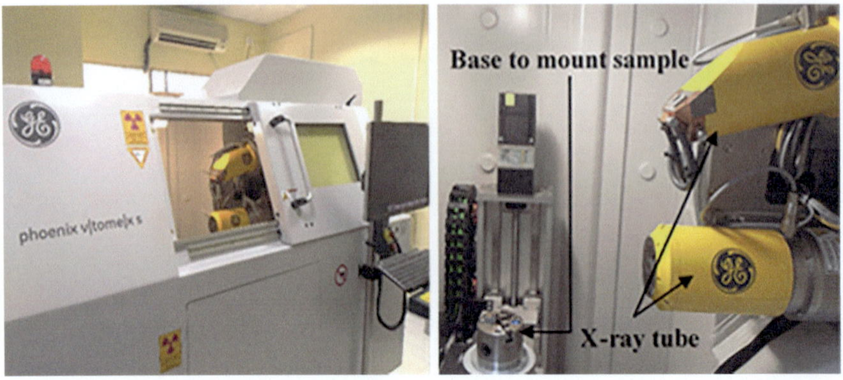

Fig. 7.5 X-ray micro tomography instrument

distribution of smaller sized pores. According to Du Plessis et al. (2016), the scanning resolution doubles with the smallest detectable pore size.

7.3 Degree of Hydration

The thermal analysis of concrete sample involving DTA/TG gives insights about the process of hydration. It measures the mass loss involved in the three major dissociation reactions as discussed earlier in the Sect. 7.2.1. Mass loss with increasing temperature is plotted in the thermogravimetric curves where each dissociation reaction corresponds to a specific temperature range (shown in Table 7.1). The mass of hydration products and the rate of reaction depend on the quantity of supplementary cementitious materials in concrete. The effects of curing ages (7, 28 and 90 days), fly ash percentages (20 and 30%), types of aggregates (RCA or NCA), and methods of mix design (PPM or IS) on the hydration reactions of concrete are discussed in this section. The concrete mixes considered in this study are designated as: Natural aggregate concrete (NAC), NAC with 20% fly ash (NFA20), NAC with 30% fly ash (NFA30), RAC with 20% fly ash (RFA20), and RAC with 30% fly ash (RFA30).

Table 7.1 Temperature corresponding to dissociation reactions

Author	Temperature ranges (°C)		
	Dehydration (ldh)	Dehydroxylation (ldx)	Decarbonation (ldc)
Bhatty et al. (1986)	105–440	440–580	580–1000
Pane and Hansen (2005)	140–440	440–520	520–1100
Monteagudo et al. (2014)	105–430	430–530	530–1100
Present study	105–433	433–527	527–820

7.3.1 *Effects of Curing Age*

The mass loss associated with the dissociation of hydrates, portlandite (CH) and $CaCO_3$, is represented by the three components ldh, ldx, and ldc in Table 7.2 As the curing age advances from 7 to 28 and then to 90 days, the values of ldh and ldx increase in NAC. As per Eqs. 7.6 and 7.7, higher values of ldh and ldx indicate higher bound water content thereby increasing the degree of hydration (α). α increases significantly for concrete cured up to 28 days and beyond that there is slight increase for 90 days cured concrete. This observation holds true for both NAC-PPM and NAC-IS concrete. The concrete mixes (both NAC and RAC), proportioned with PPM and comprising fly ash showed an increase in the component ldh when curing age increased from 7 to 28 days, followed with a reduction at 90 days. It might be explained by the fact that, a duplex layer of C–S–H and CH (Diamond et al. 1980) formed a coating surrounding the surface of spherical fly ash particle which partially inhibited pozzolanic reactions. In the bulk matrix region, fly ash offers more nucleation sites for combining with CH, facilitating higher degree secondary reactions at later ages. In contrast, formation of hydrates increased up to 90 days in FARAC mixes designed in the IS method.

The variations in the cement content may be one of the reasons for these discrepancies in the trend of the ldh and ldx components in concrete prepared using the PPM and IS methods. Fly ash incorporated concrete constructed with PPM has lower cement quantities than cementitious materials computed using the IS method of mix proportioning. In NFA20-IS and NFA30-IS, higher cement content permitted more primary hydration, which led to higher CH and C–S–H formation at 90 days, as opposed to 7 days. Conversely, the secondary reaction was more prominent in NFA20-PPM and NFA30-PPM at longer curing ages.

Table 7.3 displays the CH content that was computed using the equivalent molar weight of H_2O loss in accordance with Eq. 7.3. The higher CH% with increasing curing age in NAC also indicates the progress of hydration. On the other hand, fly ash-based NAC proportioned in PPM resulted in a slight decrease in CH% at 90 days relative to that at 7 days. It demonstrates how CH is utilized to produce additional C–S–H through the secondary hydration. Similar results of a decreasing CH concentration after 14 days due to fly ash were previously reported by Bhatty et al. (1986). Lower hydrates content at 90-days in the interface region, as mentioned earlier, was also reflected in the lower degree of hydration (α) at 90 days in the fly ash incorporated concrete. Area under an endothermic DTA peak shows a directly proportional relation to the enthalpy of the reaction or heat of hydration. In an exothermic reaction, the heat of hydration represented the quantity of heat released per unit mass or per mole of reacting binder. Additionally, enthalpy demonstrated a direct proportionality to the quantity of chemicals undergoing reaction. Area under the CH peak was lower in the presence of fly ash as compared to the mix without it. The rate of reaction declined after 28 days.

Table 7.2 Effect of curing age on hydration reaction of concrete

Concrete mix	Curing ages (days)	Weight losses (%)			Bound water (W_b) in %	Degree of hydration (α) in %
		ldh	Ldx	ldc		
NAC-PPM	7	5.22	1.52	5.15	8.8515	38.48
	28	5.31	1.59	7.61	10.0201	43.56
	90	5.42	2.15	6.12	10.0792	43.82
NFA20-PPM	7	5.82	1.46	8.08	10.5928	46.05
	28	6.33	1.32	7.28	10.6348	46.24
	90	4.03	1.24	6.79	8.0539	35.02
NFA30-PPM	7	5.53	1.32	7.01	9.7241	42.28
	28	6	1.29	4.49	9.1309	39.69
	90	4.64	1.27	6.99	8.7759	38.16
RFA20-PPM	7	5.99	1.92	11.93	12.8013	55.66
	28	6.69	1.88	11.94	13.4654	58.54
	90	5.88	1.83	9.36	11.5476	50.21
RFA30-PPM	7	3.93	1.31	10.25	9.4425	41.05
	28	5.77	1.28	9.77	11.0557	48.07
	90	4.75	1.12	11.76	10.6916	46.48
NAC-IS	7	3.11	1.53	7.56	7.7396	33.65
	28	5	1.99	9.03	10.6923	46.49
	90	5.08	2.97	6.65	10.7765	46.85
NFA20-IS	7	3.6	1.51	8.26	8.4966	36.94
	28	4.13	1.76	10.3	10.113	43.97
	90	5.67	2.36	6.31	10.6171	46.16
NFA30-IS	7	3.79	1.47	9.53	9.1673	39.86
	28	3.85	1.6	9.25	9.2425	40.18
	90	3.92	1.75	3.06	6.9246	30.17
RFA20-IS	7	3.33	1.52	15.27	9.08	39.48
	28	6	1.46	18.92	11.25	48.91
	90	6.68	1.44	20.33	12.09	52.56
RFA30-IS	7	3.33	0.95	8.09	7.5969	33.03
	28	3.46	1.29	4.72	6.6852	29.07
	90	6.63	1.18	13.33	13.2753	57.72

7.3.2 *Effect of Fly Ash in NAC*

The mass loss in Fig. 7.6 highlights the impact of fly ash on NAC proportioned using PPM mix design. NAC-7d and NAC-28d concrete showed an increase in mass after the dehydroxylation reaction which is indicative of recrystallization. Therefore,

Table 7.3 Effect of curing age on CH% of concrete

Concrete mix	Curing ages (days)	CH (%)	Area under the dihydroxylation peak (µv·s/mg)
NAC-PPM	7	6.24	12.3
	28	6.53	14
	90	8.83	14.93
NFA20-PPM	7	6.00	7.8
	28	5.43	3.47
	90	5.09	2.08
NFA30-PPM	7	5.43	3.89
	28	5.30	3.73
	90	5.22	0.92
RFA20-PPM	7	7.89	4.18
	28	7.73	2.5
	90	7.52	2.12
RFA30-PPM	7	5.39	6.13
	28	5.26	1.99
	90	4.60	0.61
NAC-IS	7	6.29	5.16
	28	8.18	8.6
	90	12.21	8.6
NFA20-IS	7	6.21	7.92
	28	7.24	0.96
	90	9.70	12.3
NFA30-IS	7	6.04	1.67
	28	6.58	3.76
	90	7.19	4.9
RFA20-IS	7	6.25	1.8
	28	6.00	2.21
	90	5.92	1.03
RFA30-IS	7	3.91	2.43
	28	5.30	2.11
	90	4.85	0.61

NAC-7d had a lower final mass decrease. The increased degree of hydration of NFA20 and NFA30 concrete in comparison to NAC also demonstrated this effect. Previous studies by Berry et al. (1994) and Lam et al. (2000) also documented this improved cement hydration in fly ash-based concrete. It was explained by the fact that the action of cement is diluted when fly ash is substituted for cement at a lower w/b ratio, which raises the effective w/c ratio. The triple mixing method formed a coating of fly ash slurry on the aggregate surface; this may have allowed for more chemically combined

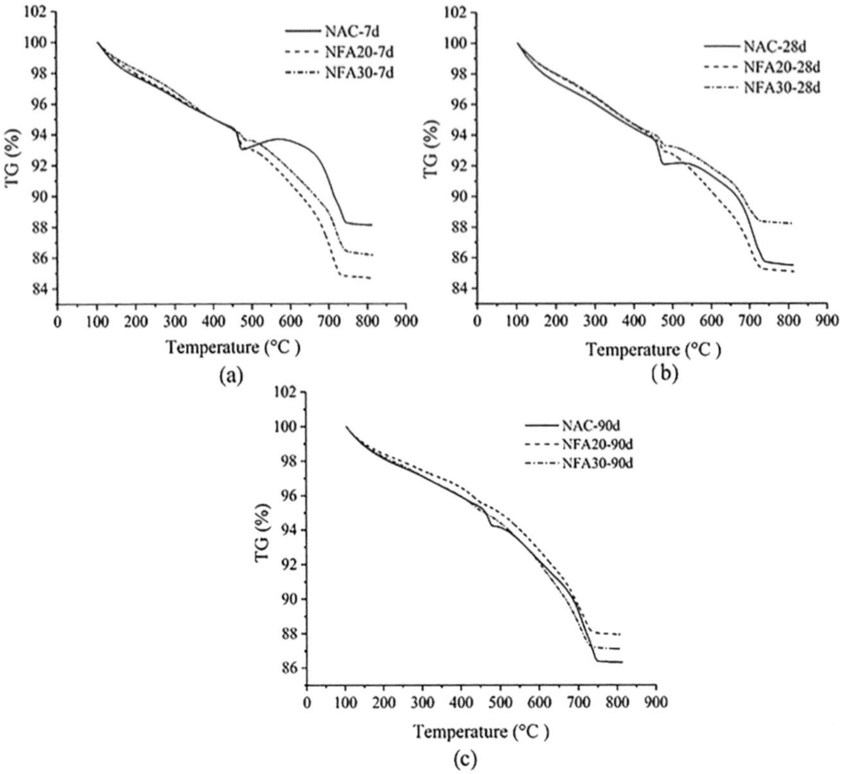

Fig. 7.6 Effect of fly ash in NAC-PPM **a** 7 days, **b** 28 days, **c** 90 days

water in NFA20 and NFA30 as opposed to NAC in PPM. Conversely, NFA20 and NFA30 concrete proportioned in IS technique demonstrated lower water-to-binder (w/b) ratio compared to NAC because more cement was available and the dilution impact was less pronounced.

Figure 7.7 illustrates the thermogravimetry curves showing mass losses in NAC mixes with and without fly ash which are proportioned using the IS mix design method. When comparing the fly ash integrated concrete to blended cement pastes (20 and 35% fly ash) in earlier research (Montegudo et al. 2014), a higher carbonation component (ldc) was reported by Sunayana and Barai (2021) which was pronounced at early ages of 7 and 28 days. This may be because the pozzolanic reaction is delayed (Sisomphon and Franke 2007) and the availability of CH is greater in the vicinity of the interface region at early ages. Regardless of the curing ages, CH% dropped in NFA20-PPM and NFA30-PPM as compared to NAC. The fact that NFA20 and NFA30 had lower total CH% suggested two things: (a) a pozzolanic reaction involving fly ash, and (b) a reduced cement content used in the mix. Greater carbonation resistance in FARAC was evidenced by a lower CH content. Fly ash in the

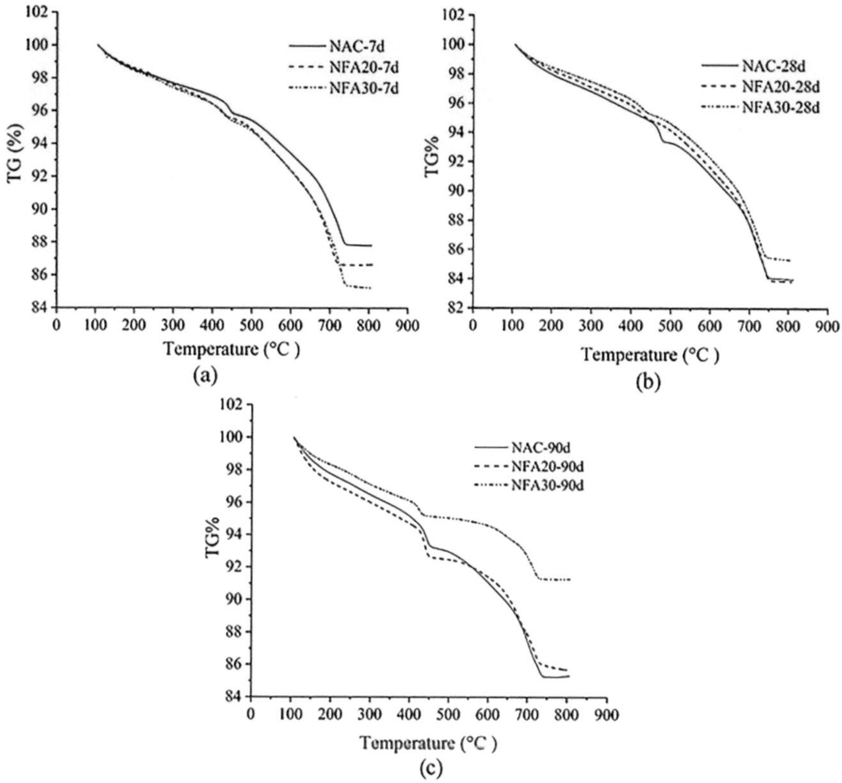

Fig. 7.7 Effect of fly ash in NAC-IS **a** 7 days, **b** 28 days, **c** 90 days

concrete caused a reduction in the dehydroxylation peak area, which also indicated a lower heat of reaction.

7.3.3 *Effect of Aggregate Types*

The effect of aggregate types (RCA or NCA) on the 20% and 30% fly ash substituted concrete proportioned in PPM is depicted in Fig. 7.8. In spite of the mix design, RFA20 and RFA30 concrete had higher total chemically bound water content and degree of hydration when compared to NFA20 and NFA30 concrete (Table 7.2). 30% fly ash integrated RAC showed increased CH consumption, which is suggestive of more pronounced secondary reactions. The triple mixing method was effective in uniformly distributing the fly ash particles around RCA.

Higher levels of hydration in the vicinity of the interface were made possible by better packing and a decreased likelihood of shadow zones in the comparatively rounder RCA as opposed to the angular NCA. At seven days, RFA20 concrete

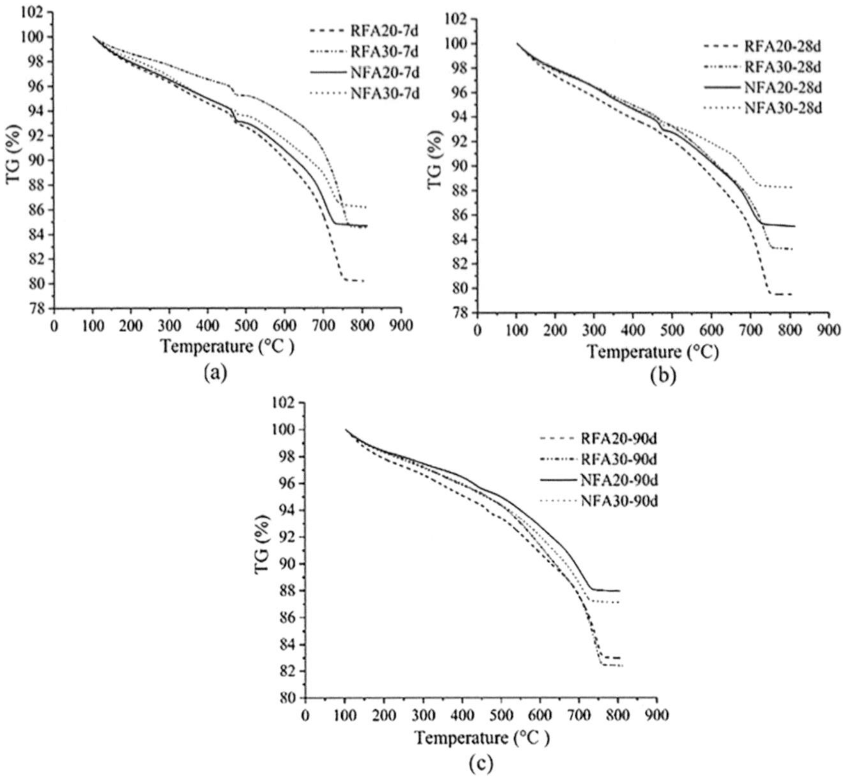

Fig. 7.8 Effect of RCA **a** 7 days, **b** 28 days, **c** 90 days

had a significantly greater CH% than NFA20, which could have been caused by the attached mortar content that was already there on the RCA surface (with some partially hydrated products or CH). The fact that RCA is made from construction wastes, that has been left on the ground after demolition of structures at the end of its useful life, may be the cause of the higher carbonation in RCA. Thus, the carbonation process is facilitated by the CH particles, if any, that are absorbed from the atmosphere by the partially hydrated adhering mortar. Total chemically bound water loss was higher due to RAC regardless of the age of hydration. In addition, replacement of RCA in concrete caused smaller proportional enthalpy of dehydroxylation reaction. The microstructure characteristics of fly ash-incorporated, 100% recycled coarse aggregate concrete are influenced by the distribution of hydration products along the interface in addition to the degree of hydration. This properties are observed using a scanning electron microscopy, the observations are discussed in the following section.

7.4 Distribution of Hydration Products

The temperature, mechanism of creation, and the amount of space available for formation all affect the morphology of hydration products. Figures 7.9, 7.10 and 7.11 represent SEM images underlining distribution of hydration products in FARAC and NAC.

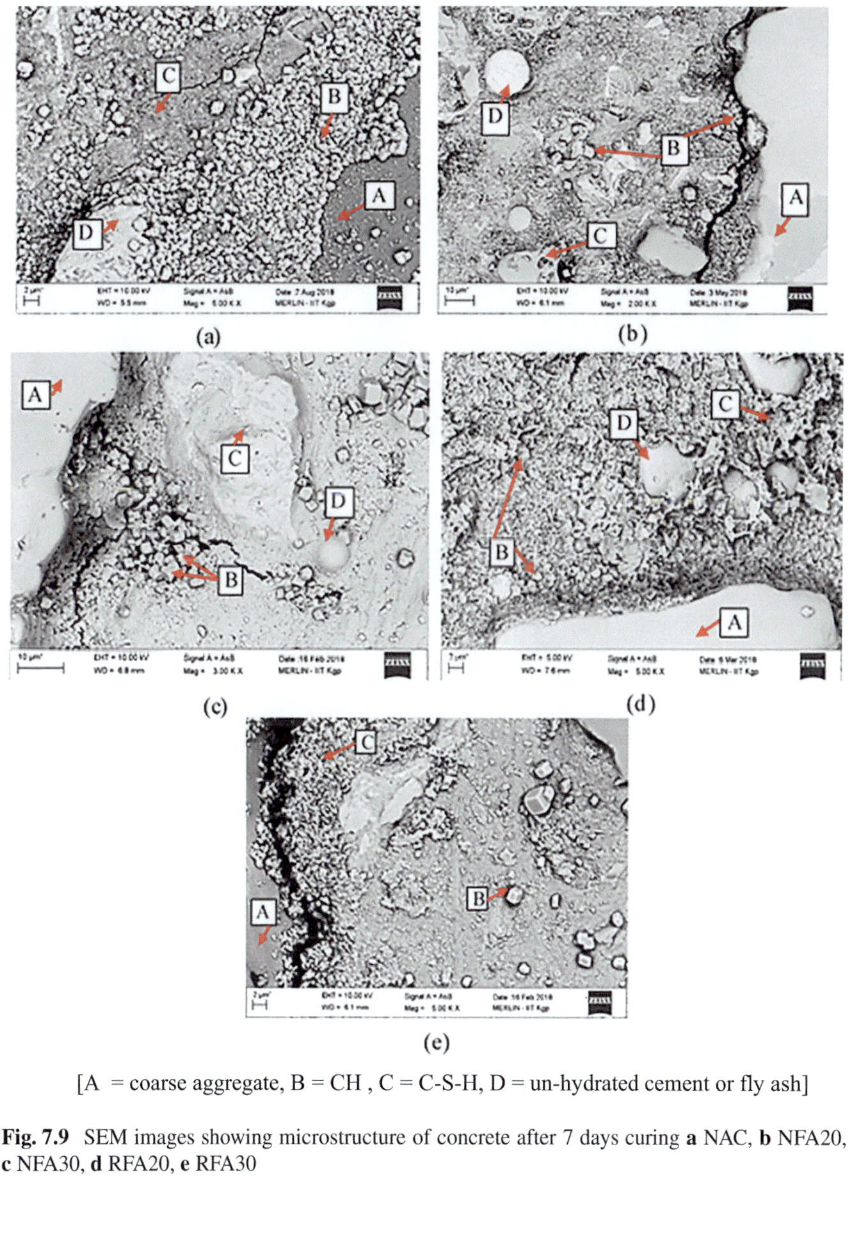

[A = coarse aggregate, B = CH , C = C-S-H, D = un-hydrated cement or fly ash]

Fig. 7.9 SEM images showing microstructure of concrete after 7 days curing **a** NAC, **b** NFA20, **c** NFA30, **d** RFA20, **e** RFA30

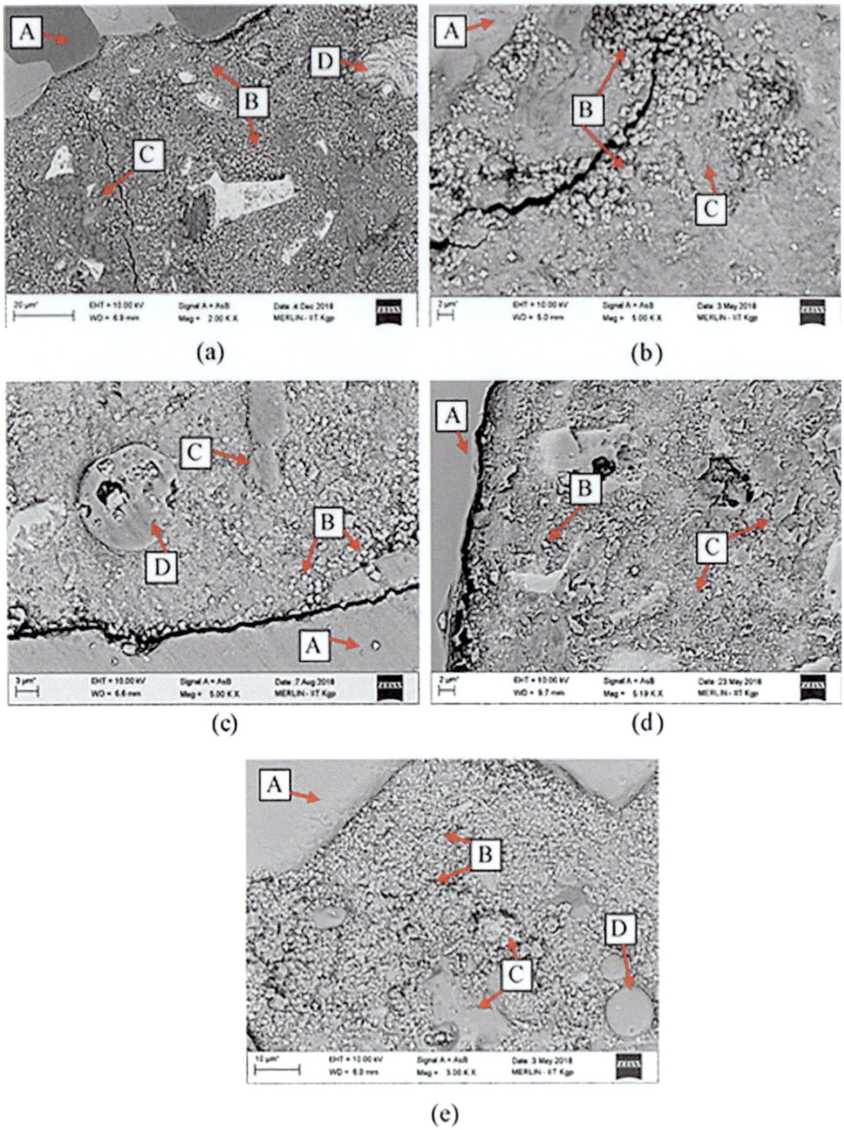

Fig. 7.10 SEM images showing microstructure of concrete after 28 days curing **a** NAC, **b** NFA20, **c** NFA30, **d** RFA20, **e** RFA30

Hexagonal calcium hydroxide (CH) crystals are present in a preferentially oriented manner in the vicinity of the interface. The bulk paste exhibited a more uniform zone of hydrated paste with fewer CH crystals. In contrast to NAC, which exhibited more CH particles, NFA20 and NFA30 concrete mixes displayed lesser CH content (quantitatively measured in the following section) and densification close to the aggregate

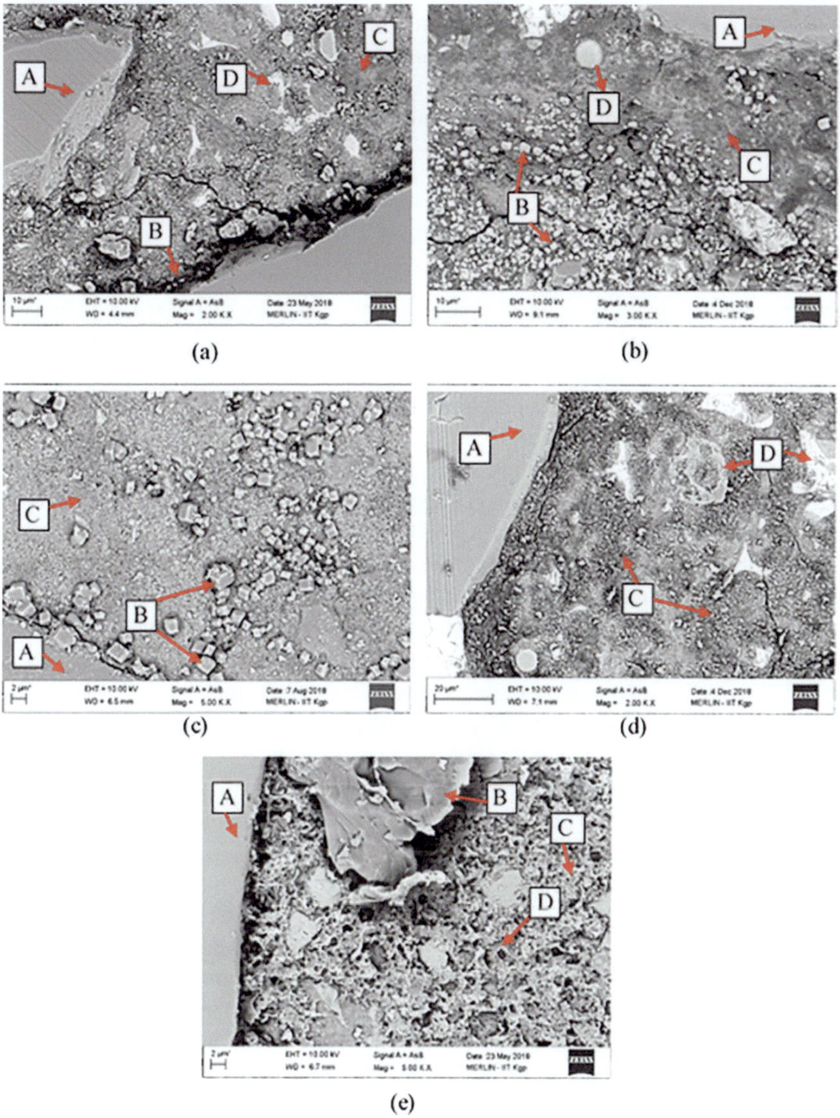

Fig. 7.11 SEM images showing microstructure of concrete after 90 days curing **a** NAC, **b** NFA20, **c** NFA30, **d** RFA20, **e** RFA30

edge. However, scanning of the RFA20 and RFA30 concrete mixes revealed CH and some sparsely connected low density calcium-silicate-hydrate (C–S–H) particles near the aggregate boundary. The presence of water-filled spaces in the interface created an optimal environment for the development of low-density C–S–H. Pozzolanic cement paste also showed an intimate mixture of nano-metre scale CH and low-density C–S–H; isolating these two phases is quite challenging (Richardson et al. 1994). In contrast, a higher concentration of C–S–H is more likely to occur in the main paste matrix where there is a greater abundance of cement particles. The old attached mortar layer in the interface of the RFA20 and RFA30 concrete specimens resulted in enhanced porosity and a few unhydrated particles.

7.4.1 Porosity, CH, and Unhydrated Phase Content (%)

The quantitative analysis of SEM images enable how various phases such as void spaces, portlandite and unreacted cement or fly ash are distributed within the hydrated concrete. The results are displayed in Figs. 7.12 and 7.13. For both RAC and NAC mixes (with or without fly ash), higher levels of porosity (%) were observed close to the aggregate-mortar interface. In NAC mixes containing fly ash or without it, as we move away from the aggregate edge, porosity dropped up to 45 μm from the interface, with a minimal variation beyond this distance. The interfacial porosity of fly ash was found to be higher in FARAC than that of NAC. After a 90-day curing period, the porosity (%) determined from the image analysis indicated a minimum value close to the interface (represented at 5 μm), thereafter it grew up to 15 μm distance from the interface. Beyond this point, porosity declined with increasing distance from the aggregate edge, suggesting that fine fly ash particles in the cement matrix improved packing of the paste thereby enhancing the interface microstructure. Additionally, the triple mixing method ensure that the recycled concrete aggregate surface was coated with fly ash. In contrast to porosity, the distribution of portlandite (CH) was relatively uniform (lesser variation with distance) across the interface, with a peak concentration at 15 μm from the aggregate edge. In contrast to the NAC mix, lower CH percentages were formed in NFA20 and NFA30 mixes, indicating that pozzolanic reactions occurred with fly ash at both 28 and 90 days, aligning with the results of DTA/TGA analysis (Table 7.3). Nevertheless, the CH percentage differed between the two analyses, with the quantitative SEM image analysis yielding higher values. TGA, conducted using powder samples collected from the aggregate edge in hardened concrete, gives a measure of average CH values at the interface. However, development of CH and their distribution around the aggregates are found to be non-uniform and localized which are captured by SEM images.

Figure 7.13 represents the distribution of unhydrated (UH) products (%) surrounding the aggregate. In the regions closer to the interface, the degree of hydration is greater, resulting in a lower UH %. NAC demonstrated a distinct trend of increasing UH percentage as the distance from the aggregate edge increased. In comparison to the reference NAC, the interface region in the fly ash-incorporated RAC and NAC mixes exhibited fewer unhydrated products. Significantly lesser UH in NFA30-28d indicated possibilities of better packing and more nucleation sites facilitating growth of hydration products. Except for RFA30-28d, FARAC displayed a lower UH% as compared to the corresponding NAC with fly ash, supporting the earlier finding that RAC had a higher degree of hydration (Table 7.2). This might be

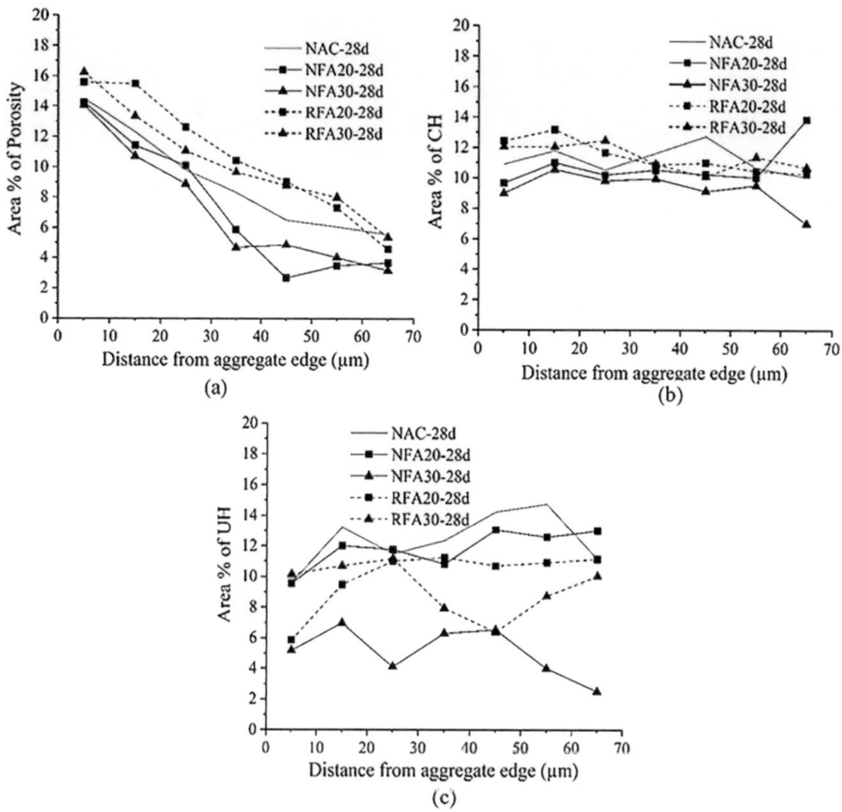

Fig. 7.12 Distribution profile of **a** porosity, **b** CH, and **c** UH (%) at 28-days

due to predominant effect of triple mixing method in RAC. Fly ash can have a beneficial effect by reducing the amount of unhydrated material, even if porosity increases when RCA is present. The hydration might have been caused by partially hydrated particles in the adhering mortar material. It is important to note that, the aggregate-mortar interface region is extremely random, localized, and complex. Consequently, to obtain a more comprehensive image of ITZ in RAC and NAC integrating fly ash, statistical variations with greater representative size and number of interface regions needs to be investigated.

7.5 Hardness and Modulus of Interface

The way the hydration products are distributed surrounding the aggregate can influence the micromechanical characteristics of the interface. To investigate these properties a length scale ranging from 10^{-6} to 10^{-8} m, where physical chemistry correlates

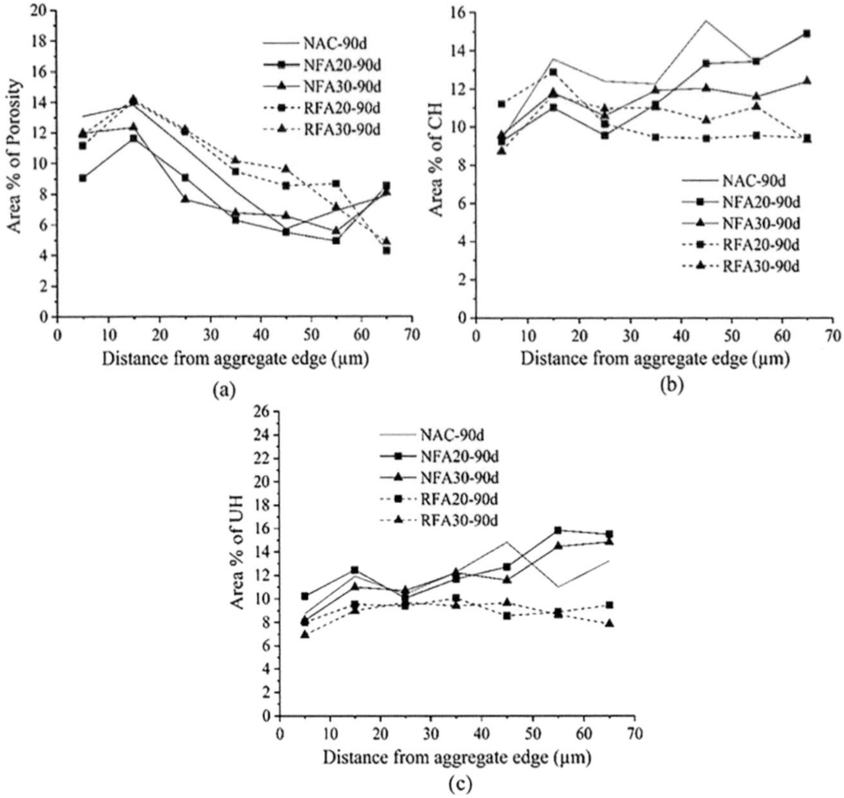

Fig. 7.13 Distribution profile of **a** porosity, **b** CH and **c** UH (%) at 90 days

well with mechanics, is appropriate (Constantinedes et al. 2003). Figure 7.14 represents the old and new interface of FARAC observed under an optical microscope at 90 days.

The hardness profile of FARAC and NAC are represented in Figs. 7.15, 7.16, 7.17 and 7.18. The interface hardness values of NAC declined up to 70 μm from the aggregate edge, reaching a minimum of 0.24 GPa. In contrast to NAC, the hardness value increased in NFA20, suggested that incorporating 20% fly ash to NAC enhanced the aggregate-mortar interface. The NFA30 displayed a substantially lower degree of fluctuation in hardness along with higher average hardness as the distance from the aggregate edge increased, with a minimum value of 5 GPa, measured at a distance of 40 μm. In RFA20, the aggregate-old mortar interface (old ITZ) showed a steady decrease in hardness, reaching 0.5 GPa at a distance of 55 μm. However, the interface between old and fresh mortar (new ITZ) lacked a consistent hardness pattern. Compared to the other concrete samples, RFA30 exhibited minimal variation in hardness along the interface, where the hardness steadily increased up to a distance of 85 μm from the edge of the aggregate.

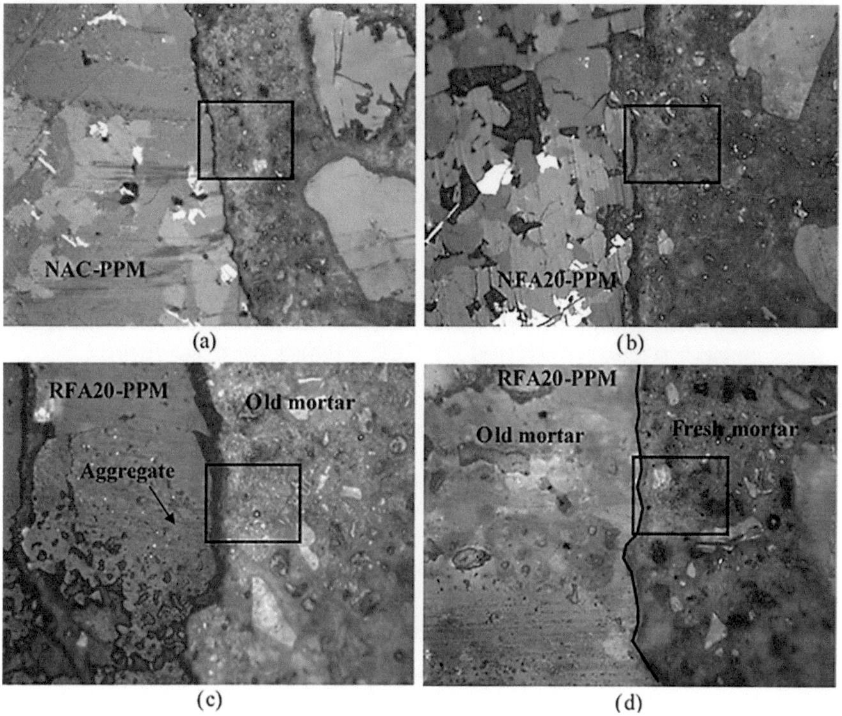

Fig. 7.14 Aggregate-matrix interface observed under optical microscope **a** NAC, **b** NFA20, **c** RFA20-old ITZ **d** RFA20 new ITZ

Fig. 7.15 Distribution of hardness in the aggregate-matrix interface (NAC PPM)

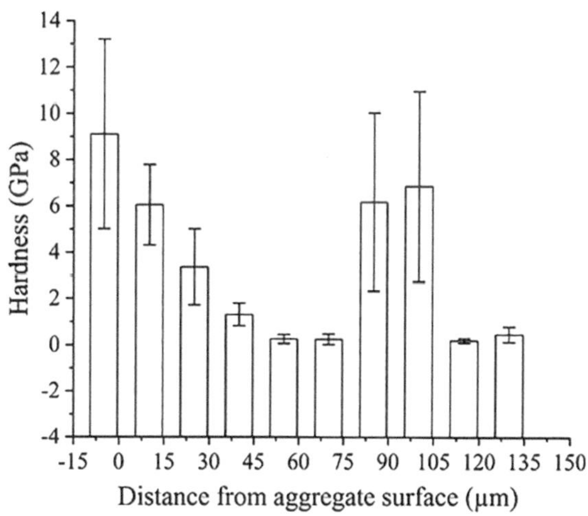

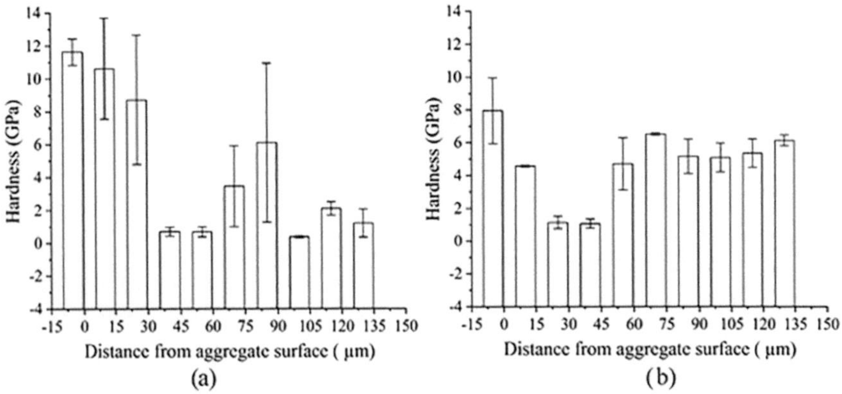

Fig. 7.16 Distribution of hardness **a** NFA20PPM and **b** NFA30PPM

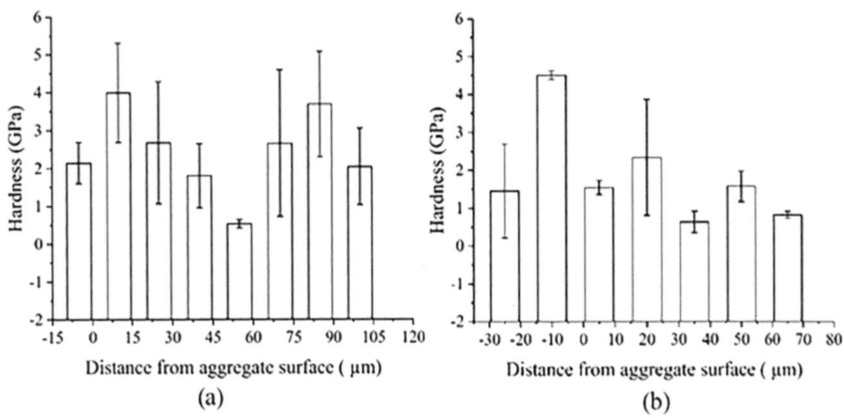

Fig. 7.17 Distribution of hardness in the **a** aggregate-matrix interface and **b** old-new mortar matrix interface (RFA20PPM)

The thickness of ITZ can be assessed as that distance from the aggregate edge beyond which the variation of hardness with distance is minimal (Xiao et al. 2013). In NAC, the ITZ thickness ranged from 50 to 70 μm, while in NFA20 and NFA30 mixes, the ITZ thickness was 40–55 μm. The old ITZ in RFA20 and RFA30 was approximately 50 μm thick, falling within the range of NAC. The interface hardness profile indicated reduced hardness in case of RFA20 and RFA30 mixes.

Elastic modulus in the ITZ region of concrete is illustrated in Figs. 7.19, 7.20, 7.21 and 7.22. The lowest modulus value of 10.17 GPa was observed at 70 μm from the aggregate edge in NAC. In case of NFA20 concrete, a minimum modulus of 13.04 GPa was reported at 55 μm.

NFA30 concrete exhibited comparatively greater elastic modulus at the interface, likely due to the availability of higher unhydrated cement content as evidenced from

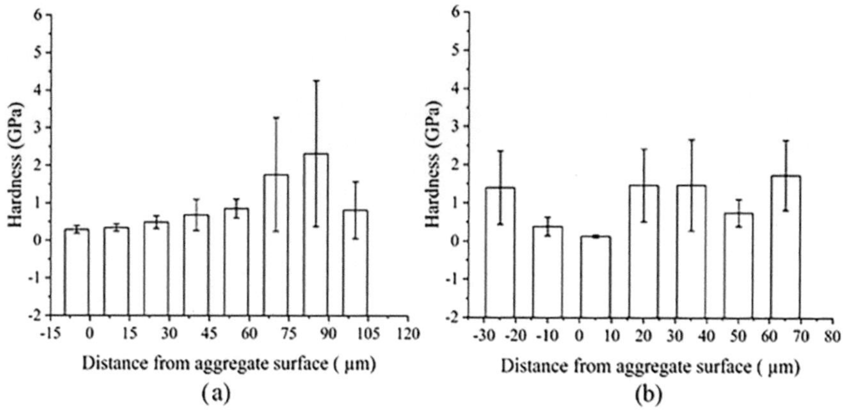

Fig. 7.18 Hardness distribution in the **a** aggregate-matrix interface and **b** old-new mortar matrix interface (RFA30PPM)

Fig. 7.19 Distribution of modulus in the aggregate-matrix interface (NACPPM)

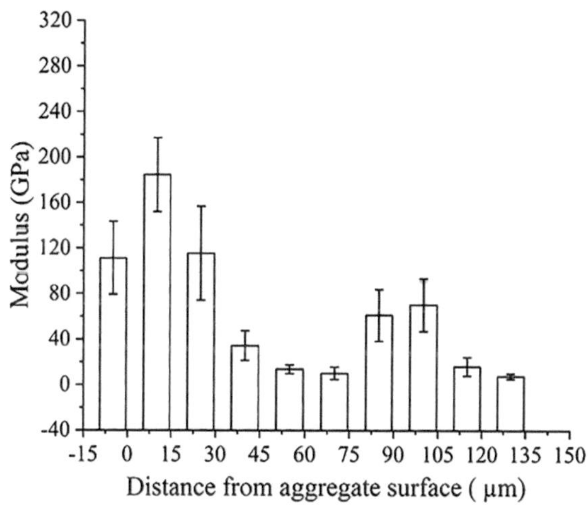

the SEM images previously shown. The incorporation of fly ash as a cement substitute enhanced ITZ properties. Different phases in a hydrated paste can be characterized by their respective elastic modulus values in an increasing order as follows: less than 7 GPa for pores, 7–34 GPa for other hydration products, 34–50 GPa corresponds to CH and unhydrated cement showing more than 50GPa (Sorelli et al. 2008; Němeček et al. 2011). In RFA20, the old ITZ modulus was minimum (16 GPa) at 55 μm from the aggregate edge, while the modulus distribution in the new ITZ showed no consistent pattern. The minimum average modulus within the interface between the old and new paste matrix was 34 GPa, observed at 65 μm from the boundary. A smaller variation in modulus across the new ITZ indicates reduced heterogeneity. In RFA30 concrete, modulus in the old interface was higher than that in the new interface between the

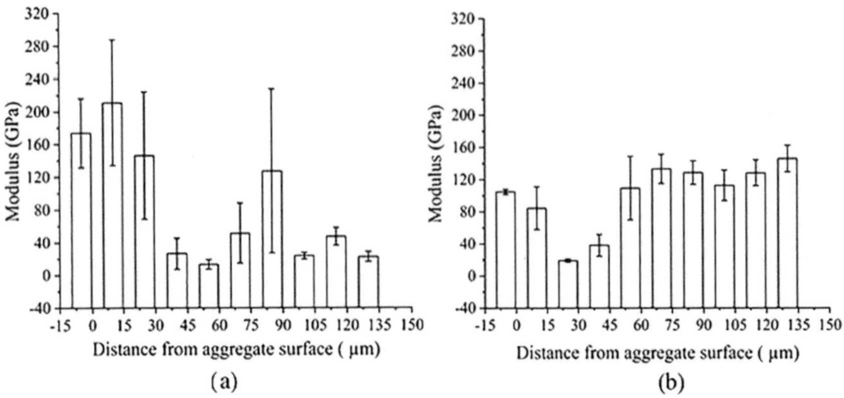

Fig. 7.20 Distribution of modulus in **a** NFA20PPM and **b** NFA30PPM

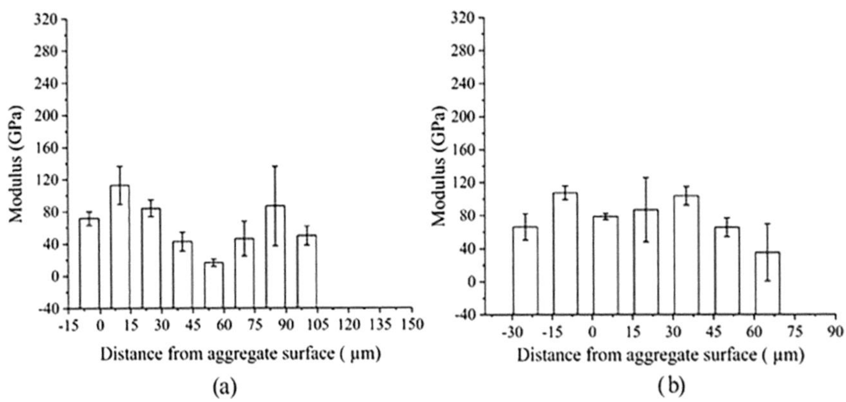

Fig. 7.21 Distribution of modulus in RFA20PPM **a** aggregate-matrix interface, **b** old-new mortar matrix interface

old and fresh mortar. The new ITZ displayed the lowest modulus value of 4 GPa at 35 μm from the aggregate edge. The cause for this reduced elastic modulus of the new ITZ in RFA30 is not clear warranting the further investigation.

7.6 Distribution of Pores in the Specimens in Three Dimension

Three-dimensional non-destructive scanning of 100 mm diameter and 200 mm height cylinder specimens were conducted to visualize distribution of pores within the hardened concrete (Sunayana and Barai 2021). Figure 7.23 represents the scanned images showing concrete slices and pores distribution throughout the whole specimens for

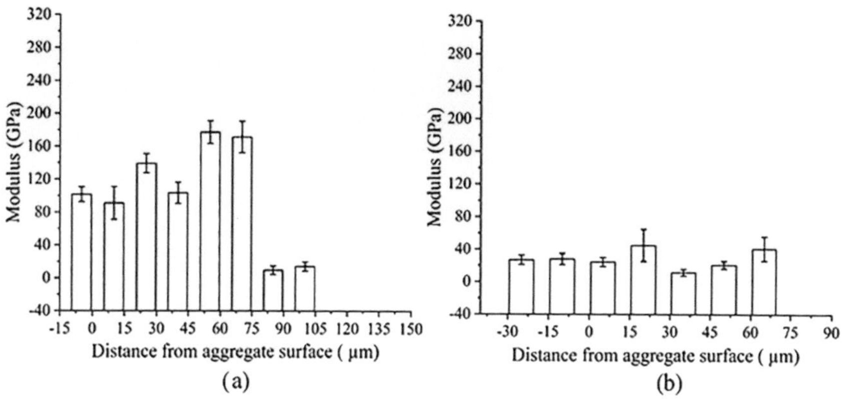

Fig. 7.22 Modulus distribution in the **a** aggregate-matrix interface and **b** old-new mortar matrix interface of RFA30PPM

typical NFA20 and RFA20 concrete mixes. Numerous discrete small size pores were found in the concrete specimen constituting of NAC with fly ash. In contrast, in RFA20 specimen, a combination of few discrete pores as well as some connected pores are observed. Figure 7.24 displays the distributions of different sizes of pores in the cylinder specimens cured up to 7, 28 and 90-days, with pore diameter indicating equivalent spherical diameter. Partially substituting cement with fly ash in both NAC and RAC led to formation of higher number of smaller size voids. The influence of fly ash on pore refinement was notably enhanced in the presence of recycled concrete aggregates, refining pores up to 0.6 mm, independent of any curing duration.

7 days-cured RFA20 mix showed higher porosity (1.05%) as well as mean pore diameter (0.84 mm) than NFA20 which exhibited porosity of 0.5% and mean pore diameter of 0.81 mm. In RFA20, the individual pore volume ranged from 0.02 to 30.69 mm^3 (mean pore volume = 0.125 mm^3), while in NFA20, they ranged from 0.02 to 26.88 mm^3 (mean pore volume = 0.138 mm^3). As the percentage fly ash in RAC increase, porosity, mean pore diameter and pore volume also increase. RFA30 specimen showed 1.22% porosity with mean pore diameter 0.86 mm and mean pore volume of 0.217 mm^3. For 28 days cured concrete specimens, individual pore volumes in NFA20 (mean volume = 0.107 mm^3) and RFA20 (mean volume = 0.129 mm^3) ranged between 0.02 and 24 mm^3, resulting in porosities of 0.71% (mean pore diameter = 0.82 mm) and 1.03% (mean pore diameter = 0.83 mm) respectively. In contrast, the pore volume in NFA30 and RFA30 varied between 0.02–27 mm^3 (mean volume = 0.197 mm^3) and 0.02–34 mm^3 (mean volume = 0.215 mm^3) respectively. The corresponding porosities values were 0.63% (mean pore diameter = 0.82 mm) and 0.7% (mean pore diameter = 0.82 mm) respectively in NFA30 and RFA30. As hydration continued, the porosity, mean pore diameter and individual pore volume decreased. By 28 days, the impact of RAC on porosity was minimal, as indicated by the diminished differences between NAC and RAC incorporating fly ash, supporting similar 28-day compressive strengths across aggregate types.

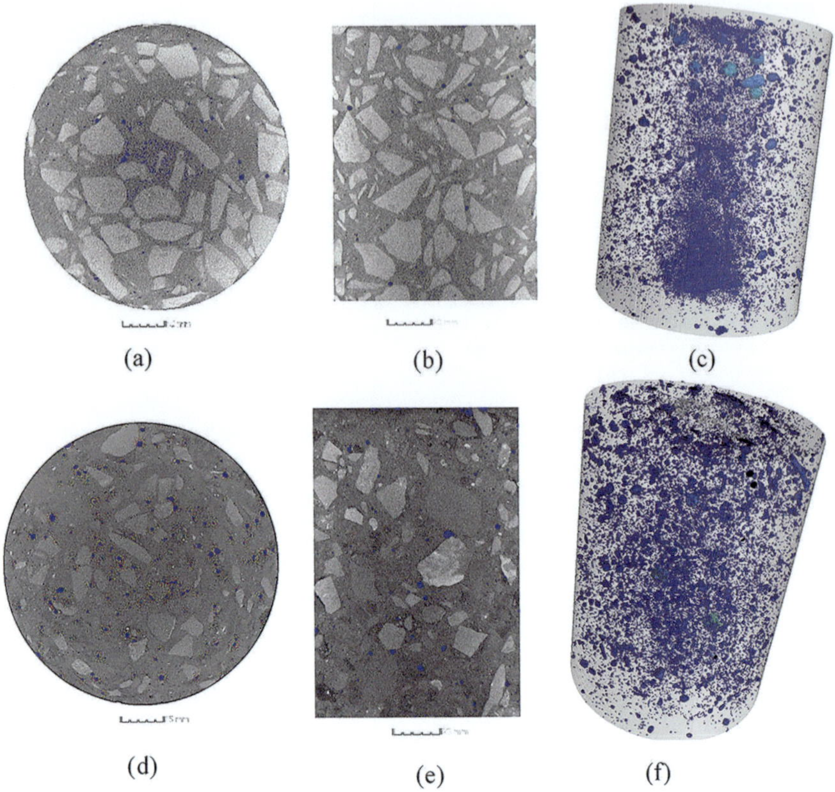

Fig. 7.23 X-ray computed tomographic images **a–c** NFA20, **d–f** RFA20

Following 90 days curing period, the porosity of NFA20 was measured as 0.64% (mean pore diameter = 0.83 mm), while RFA20 exhibited a porosity of 0.56% (mean pore diameter = 0.79 mm). NFA30 and RFA30 displayed nearly comparable porosity of 0.7%, though RFA30 showed a higher maximum pore volume (30.42 mm^3), possibly due to the presence of micro cracks. The increased porosity associated with RCA can be mitigated due to development of larger number of smaller size pores when fly ash is present. The occurrence of larger air-voids remained largely unaffected by the type of concrete. However, RFA20 and RFA30 concrete demonstrated a lower relative frequency of these larger voids.

Sphericity, which refers how closely a shape resembles that of a sphere is a key parameter used for characterization of pores. The sphericity value of a perfect sphere is quantified as 1. The sphericity value of voids ranging from 0.35 to 4 mm size are compared in Fig. 7.25. The figure illustrates voids developed in concrete specimens prepared using PPM and IS methods of mix design and tested after 7, 28 and 90 days of curing. Voids having sphericity values above 0.5 suggest a tendency towards formation of more spherical pores, independent of concrete types. The inclusion of fly

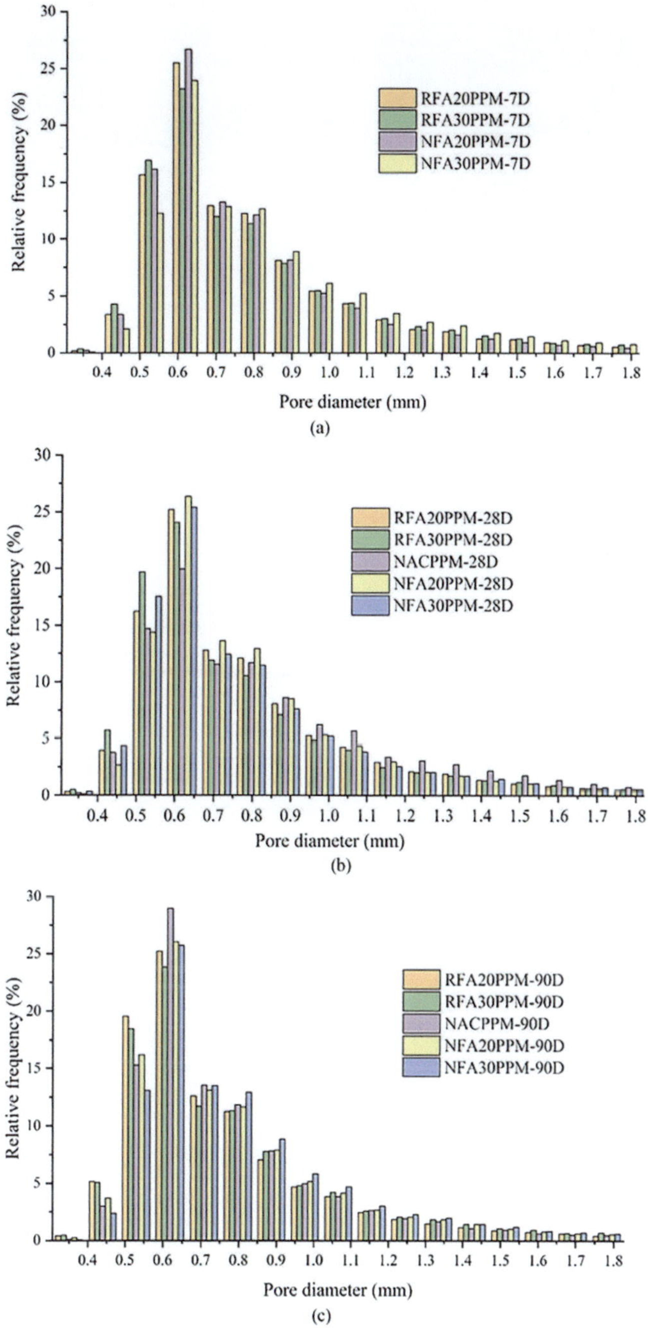

Fig. 7.24 Pore size distribution in fly ash based NAC and RAC in cylinder sample cured at **a** 7 days, **b** 28 days and **c** 90 days

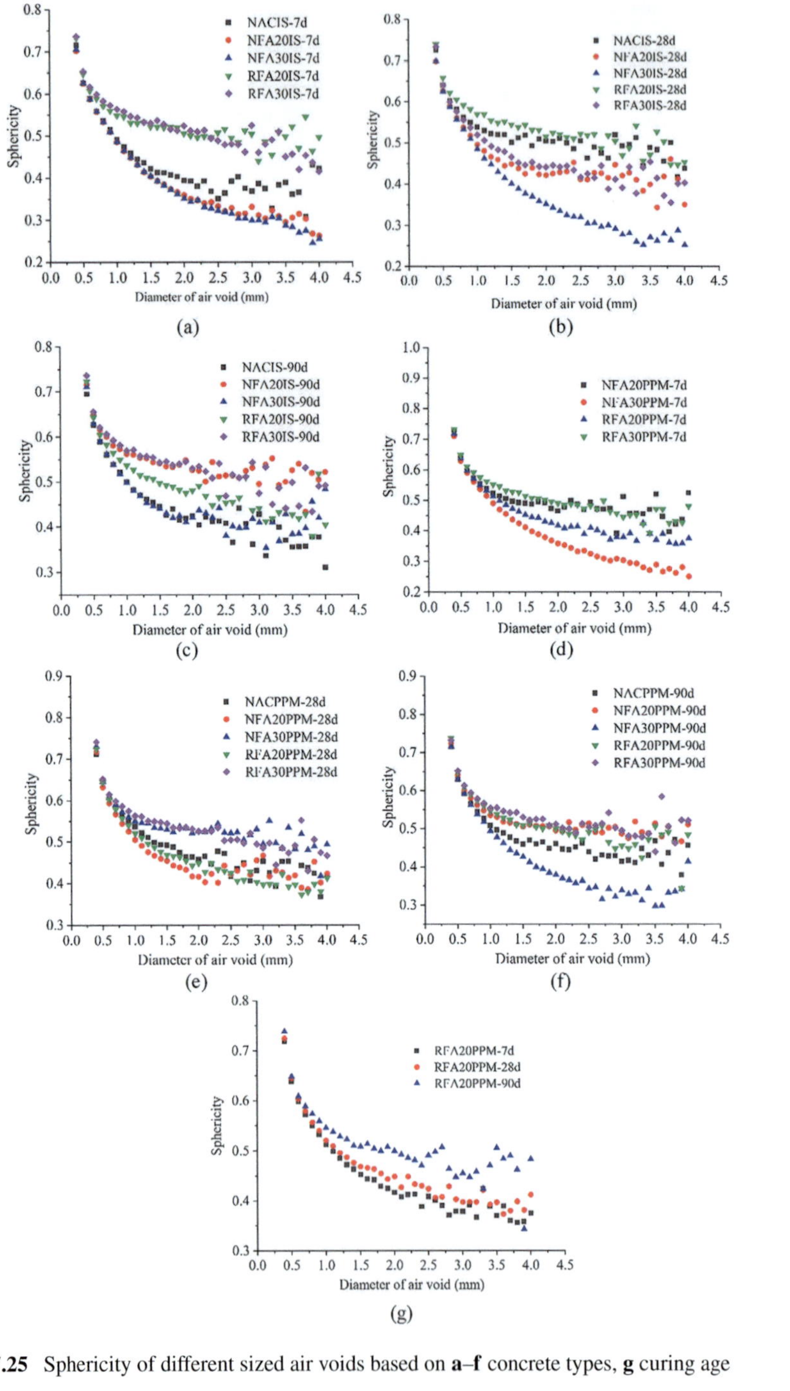

Fig. 7.25 Sphericity of different sized air voids based on **a**–**f** concrete types, **g** curing age

ash in concrete led to a greater number of near-spherical air voids; especially in case of RAC with fly ash. Formation of more spherical pores suggests potentially lower pore connectivity. However, further studies are needed to evaluate actual connectivity of pores. Smaller voids exhibited higher sphericity compared to larger ones, indicating that the shape of smaller voids tends to be more spherical. Increasing the fly ash content raised sphericity of voids in both NAC and RAC suggesting that fly ash positively contribute towards development of more spherical shaped voids in RCA replaced concrete. For example: Fig. 7.25g explains the effect of curing age on sphericity of air voids in RFA20. With the advancement of hydration during curing, there was a reduction in porosity and an increase in sphericity of pores, indicative of a well distributed spherical pores over time. Figure 7.26 represents pore distributions in smaller cuboid samples extracted from the cylinder specimens that were cured for 7, 28 and 90 days. Smaller pores, within the 25–350 μm range, were more numerous in NFA20 and NFA30, while larger pores appeared in RFA20 and RFA30, likely due to microcracks and adhered mortar in RAC.

7.7 Relation Between Micro- and Macro-level Properties

The complexity in translating micro-level behaviour to macro-level response in concrete arises from the high heterogeneity at the interface, the interactions among hydration reaction, rate of hydration, morphology, distribution of hydrated products, and porosity. Nevertheless, this study attempts to provide some insights into behaviour of recycled coarse aggregate concrete with fly ash based on its microstructural properties.

Figure 7.27 illustrates the correlation between total porosity and compressive strength across different concrete mix considered in this study. The curve fitting utilized an exponential function which demonstrated a strong correlation between compressive strength and porosity. Notably, as discussed previously, fly ash substitution at low-volume can offset relatively high porosity of RAC by forming a higher number of spherical and smaller size pores. Extended curing periods also show higher propensity towards forming spherical pores, which may help regulate pore connectivity—a factor potentially affecting concrete durability. However, further research is needed to verify the exact relationship between pore shape and connectivity.

The degree of hydration is a crucial factor influencing both porosity and concrete strength, although the distribution of hydrated products around the aggregates is also highly significant. For RFA20 and RFA30 concrete, the degree of hydration calculated at interface was higher than fly ash incorporated NAC. However, this did not directly translate into an increased strength of FARAC, potentially due to the formation of more loosely connected low-density C–S–H at the interface when fly ash is present. The decrease in CH content indicated substantial pozzolanic activity of fly ash following 7-days curing period. The thickness of old and new ITZ, derived from nano-indentation testing, was comparable in FARAC and NAC, contrary to previous literatures that reported higher thickness in RAC. This suggests that ITZ

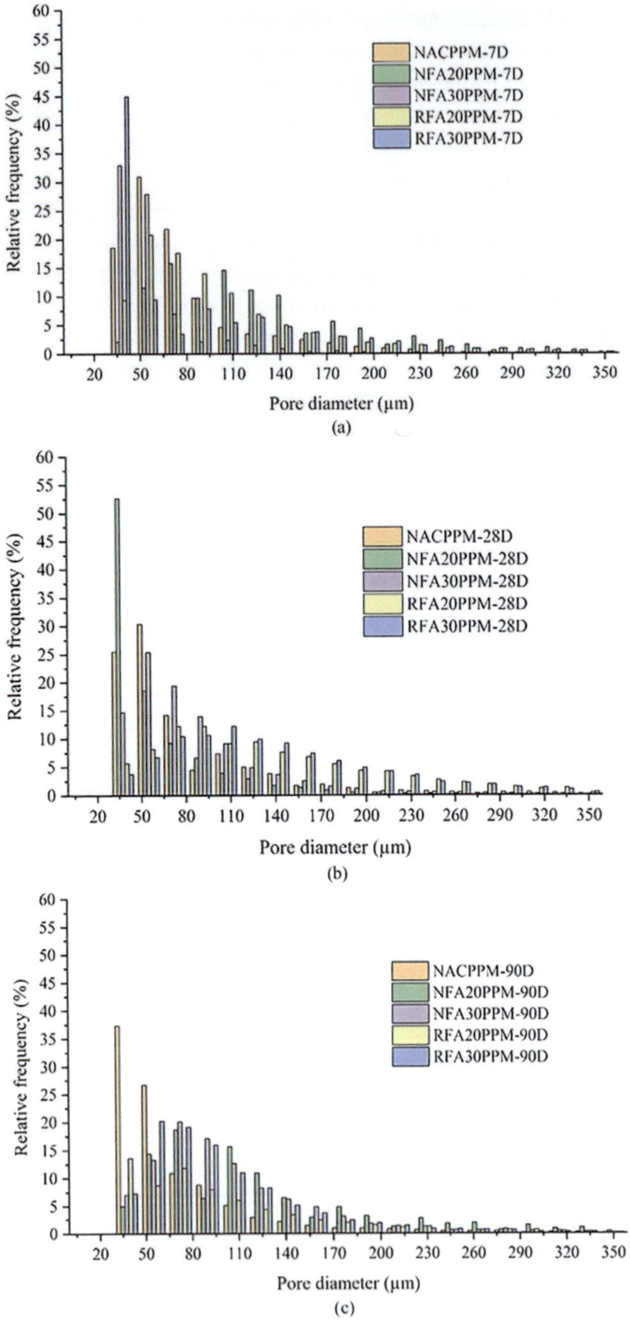

Fig. 7.26 Pore size distribution in fly ash-based NAC and RAC in cuboid sample cured at **a** 7 days, **b** 28 days and **c** 90 days

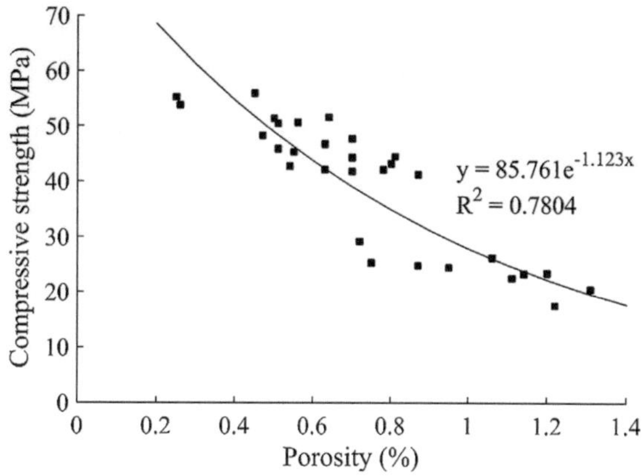

Fig. 7.27 Compressive strength versus porosity for fly ash incorporated RAC and NAC

may be modified in the presence of fly ash, likely due to finer particle sizes which facilitate better packing, triple mixing method and pozzolanic reactions that lower the presence of preferentially oriented CH in the interface region. In addition to the compressive strength, the characteristics of ITZ also affect the ratio of compressive to tensile strength of concrete (Mehta and Monteiro 2006). This ratio declined as compressive strength increased in both NAC and RAC; however, beyond 25 MPa in RAC, the rate of tensile strength gain decreased more rapidly as compared to NAC (Sunayana and Barai 2017).

7.8 Effect of Other Cementitious Materials on the Microstructure of RAC

Silica fume (SF) used in 100% RAC as a SCM can make the ITZ denser and bridge some of the microcracks (Jalilifar and Sajedi 2021). They have used SEM technique to examine the microstructure of concrete and reported similar density and hydration products in the ITZ region of both RAC with 5 and 10% silica fume and NAC. The coating of SF around RCA formed during the triple mixing method is beneficial in reducing the crystalline products like portlandite around RCA. Zhu et al. (2022) also reported positive effects of SF in RAC by qualitatively analysing the microstructure of concrete using SEM and energy dispersive spectroscopy (EDS). Treating recycled aggregate with the silica fume strengthens the RCA due to filling effect, pozzolanic reaction and restraining effect of the silica fume layer.

Steel slag powder (SP) and glass powder also affect the evolution of microstructure in RAC as found from the recent study by (Zhan et al. 2022). They have used the experimental techniques like SEM, XRD and mercury intrusion porosimetry (MIP). RAC with 20% SP makes the C–S–H flocculated. Due to the interconnectedness of the hydration products and lower CH, number of large pores reduces and the microstructure becomes denser. SP and glass powder used combinedly in RAC can reduce the thickness of ITZ and increase the elastic modulus as observed from the nanoindentation study.

7.9 Summary

This chapter discusses microstructure of cementitious materials incorporated recycled aggregate concrete. The results can be summarized as follows:

(1) Silica fume coating around RCA formed during the triple mixing method is beneficial in reducing the crystalline products like portlandite in the interface. The advantage of the combined application of steel slag and glass powder in RAC in terms of reducing the thickness of ITZ and increasing the elastic modulus can be observed from nanoindentation study.
(2) The initial higher carbonation component of recycled coarse aggregate, caused by the susceptibility of partially hydrated CH in the attached mortar to atmospheric CO_2 prior to recycling can be decreased by the pozzolanic reaction in fly ash incorporated RAC (FARAC). The particle packing approach of mix design promotes significant secondary reaction, unlike the conventional Indian standard method of mix design, which mainly fosters primary hydration due to a higher cement content.
(3) Improved packing, reactivity potential of some partially hydrated CH in the adhered mortar, reduced likelihood of existence of shadow zone in spherical RCA particle and the effects of internal curing contributed to a higher degree of hydration in the region closer to the interface. However, the increased degree of hydration in the interface region of FARAC did not directly yield higher compressive strength, as it led to the formation of more loosely bound, low density C–S–H.
(4) After 90 days curing in FARAC, the micro-cracks primarily traversed the old and fresh bulk paste matrix rather than the old ITZ (the interface between parent aggregate and adhered mortar layer). Additionally, a reduction in porosity and CH content was observed within 15 mm distance from the aggregate edge, suggesting a strengthening of the old ITZ due to the combined effects of PPM and triple mixing method with fly ash in RAC.
(5) The thickness of old ITZ in FARAC was measured as 55 mm which is comparable to the ITZ thickness in fly ash incorporated NAC. While the thickness of the new ITZ could not be definitively determined from the hardness and modulus distribution profiles, which showed no clear trend. However, reduced variability in the hardness and elastic modulus suggests lower heterogeneity within the new ITZ.
(6) Three-dimensional porosity analysis using micro tomography testing revealed increased porosity due to RCA, even with fly ash. However, the negative impact of RCA on porosity and pore distribution minimized by 28 days, aligning with the compressive strength of fly ash incorporated RAC and NAC. With an increase

in fly ash replacement from 20 to 30%, cumulative porosity decreased while the proportion of smaller pores contributing to porosity increased. Smaller pores exhibited a stronger inclination towards a near-spherical shape, especially with higher fly ash content. These spherical pores may have reduced connectivity, potentially enhancing durability of FARAC, though further research is necessary in this regard. Thus, incorporating low-volume fly ash up to 30% in RAC is beneficial for positively modifying the microstructure of RAC.

References

Baofan C (2013) Basic mechanical properties and microstructural analysis of recycled concrete. J Wuhan Univ Technol-Mater Sci Ed 28(1):104–109

Berry EE, Hemmings RT, Zhang MH, Cornelius BJ, Golden DM (1994) Hydration in high-volume fly ash concrete binders. ACI Mater J 91(4):382–389

Bhatty JI, Dollimore D, Gamlen GA, Mangabhai RJ, Olmez H (1986) Estimation of calcium hydroxide in OPC, OPC/PFA and OPC/PFA/polymer modified systems. Thermochim Acta 106:115–123

Brand AS (2015) Interfacial transition zone composition and bonding in cementitious materials with asphalt-coated particles. Doctoral dissertation, University of Illinois at Urbana-Champaign

Constantinides G, Ulm FJ (2004) The effect of two types of CSH on the elasticity of cement-based materials: results from nanoindentation and micromechanical modeling. Cem Concr Res 34:67–80

Constantinides G, Ulm FJ, Van Vliet K (2003) On the use of nanoindentation for cementitious materials. Mater Struct 36:191–196

Diamond S (2001) Considerations in image analysis as applied to investigations of the ITZ in concrete. Cement Concr Compos 23:171–178

Diamond S, Ravina D, Lovell J (1980) Occurrence of duplex films on flyash surfaces. Cem Concr Res 10:297–300

Du Plessis A, Olawuyi BJ, Boshoff WP, Le Roux SG (2016) Simple and fast porosity analysis of concrete using X-ray computed tomography. Mater Struct 49:553–562

Etxeberria M, Vázquez E, Mari A (2006) Microstructure analysis of hardened recycled aggregate concrete. Mag Concr Res 58(10):683–690

Jalilifar H, Sajedi F (2021) Micro-structural analysis of recycled concretes made with recycled coarse concrete aggregates. Constr Build Mater 267:121041

Koenders EAB, Pepe M, Martinelli E (2014) Compressive strength and hydration processes of concrete with recycled aggregates. Cem Concr Res 56:203–212

Kong D, Lei T, Zheng J, Ma C, Jiang J, Jiang J (2010) Effect and mechanism of surface-coating pozzalanics materials around aggregate on properties and ITZ microstructure of recycled aggregate concrete. Constr Build Mater 24(5):701–708

Lam L, Wong YL, Poon CS (2000) Degree of hydration and gel/space ratio of high-volume fly ash/cement systems. Cem Concr Res 30(5):747–756

Lee JS (1983) Digital image smoothing and the sigma filter. Comput Vision, Graphics Image Process 24(2):255–269

Leite MB, Monteiro PJM (2016) Microstructural analysis of recycled concrete using X-ray microtomography. Cem Concr Res 81:38–48

Li W, Xiao J, Sun Z, Kawashima S, Shah SP (2012) Interfacial transition zones in recycled aggregate concrete with different mixing approaches. Constr Build Mater 35:1045–1055

Marsh BK, Day RL (1988) Pozzolanic and cementitious reactions of fly ash in blended cement pastes. Cem Concr Res 18(2):301–310

Mehta PK, Monteiro P (2006) Concrete: microstructure, properties, and materials
Monteagudo SM, Moragues A, Gálvez JC, Casati MJ, Reyes E (2014) The degree of hydration assessment of blended cement pastes by differential thermal and thermogravimetric analysis. Morphological evolution of the solid phases. Thermochim Acta 592:37–51
Němeček J, Šmilauer V, Kopecký L (2011) Nanoindentation characteristics of alkali-activated aluminosilicate materials. Cement Concr Compos 33(2):163–170
Pane I, Hansen W (2005) Investigation of blended cement hydration by isothermal calorimetry and thermal analysis. Cem Concr Res 35(6):1155–1164
Richardson IG, Brough AR, Groves GW, Dobson CM (1994) The characterization of hardened alkali-activated blast-furnace slag pastes and the nature of the calcium silicate hydrate (CSH) phase. Cem Concr Res 24(5):813–829
Scrivener KL (2004) Backscattered electron imaging of cementitious microstructures: understanding and quantification. Cement Concr Compos 26(8):935–945
Shafiq N, Nuruddin F (2010) Degree of hydration of OPC and OPC/FA pastes dried in different relative humidity. Concr Res Lett 1(3)
Sisomphon K, Franke L (2007) Carbonation rates of concretes containing high volume of pozzolanic materials. Cem Concr Res 37(12):1647–1653
Sorelli L, Constantinides G, Ulm FJ, Toutlemonde F (2008) The nano-mechanical signature of ultra high performance concrete by statistical nanoindentation techniques. Cem Concr Res 38(12):1447–1456
Sunayana S, Barai SV (2017) Recycled aggregate concrete incorporating fly ash: comparative study on particle packing and conventional method. Constr Build Mater 156:376–386
Sunayana S, Barai SV (2021) Partially fly ash incorporated recycled coarse aggregate based concrete: microstructure perspectives and critical analysis. Constr Build Mater 278:122322
Tam VWY, Gao XF, Tam CM, Ng KM (2009) Physio-chemical reactions in recycle aggregate concrete. J Hazard Mater 163(2–3):823–828
Vedalakshmi R, Raj AS, Srinivasan S, Babu KG (2003) Quantification of hydrated cement products of blended cements in low and medium strength concrete using TG and DTA technique. Thermochim Acta 407(1–2):49–60
Wong HS, Head MK, Buenfeld NR (2006) Pore segmentation of cement-based materials from backscattered electron images. Cem Concr Res 36:1083–1090
Xiao J, Li W, Sun Z, Lange DA, Shah SP (2013) Properties of interfacial transition zones in recycled aggregate concrete tested by nanoindentation. Cement Concr Compos 37:276–292
Zhan PM, Zhang XX, He ZH, Shi JY, Gencel O, Yen NTH, Wang GC (2022) Strength, microstructure and nanomechanical properties of recycled aggregate concrete containing waste glass powder and steel slag powder. J Clean Prod 341:130892
Zhu L, Ning Q, Han W, Bai L (2022) Compressive strength and microstructural analysis of recycled coarse aggregate concrete treated with silica fume. Constr Build Mater 334:127453

Part III
Structural Performance Assessment

Chapter 8
Behaviour of Reinforced Concrete Beams

Preview

In the preceding chapters, we have discussed the properties of concrete, specifically focussing on the material behaviour of SCM and recycled coarse aggregates. The primary challenge in incorporating this sustainable recycled material is to fulfil the performance requirements of structural elements subjected to flexural, shear, and compression. The behaviour of reinforced concrete RAC beams under bending and shear is discussed in this chapter. Load carrying capacity, failure pattern and effects of material parameters are examined for fly ash incorporated 100% RAC beams. We evaluate the suitability of current code provisions for natural aggregate concrete to be used in RAC with fly ash beams. Furthermore, reliability of structural elements is evaluated, considering the variabilities associated with these recycled materials. Necessary modifications are suggested to achieve the same level of safety as conventional concrete.

8.1 Behaviour of RAC Beams Under Flexure

The load-carrying capacity of reinforced concrete (RC) beams under flexure loading is primarily determined by the tension reinforcement (ρ), cross-sectional dimension of beam and strength of concrete. The material stress–strain formulations cannot directly predict the flexure behaviour of RC beam because they do not account for the bond between concrete and reinforcing steel. Substituting coarse RCA in concrete may impact the behaviour of beams. This chapter examines the effects of different factors such as reinforcement percentage, mix design method, compressive strength of concrete and coarse RCA replacement percentage on the flexure behaviour of RC beams. How do the RC beams comprising 100% coarse recycled concrete aggregates and low-volume fly ash as cement substitute behave under flexure? Let's discuss the relevant studies on RAC and FARAC to answer this question.

S. Sunayana and S. V. Barai, *Green Concrete*, Springer Transactions in Civil and Environmental Engineering, https://doi.org/10.1007/978-981-96-6982-0_8

A study conducted by Ajdukiewicz and Kliszczewicz (2007) examined the flexural performance of beams made from RAC and NAC. They examined the effects of percentage of tension reinforcement ρ (0.77%, and 1.37%) and concrete strength. They recommended applying medium strength (class C35/45) and high strength concrete (class C80/95) in the production of coarse RCA. A marginal reduction of about 3.5% was observed in the flexural strength as compared to NAC beams. However, the service load deflection increased significantly by 18–100% in the recycled concrete beams. The necessity of conducting additional research on the increased long-term deflection resulting from creep and shrinkage was emphasized.

Bai and Sun (2010) investigated the influence of coarse RCA replacement at various levels (50, 70 and 100%) and different ρ ranging from 0.68, 0.89 and 1.13% on the flexure performance of beams. They found that as the percentage of RCA replacement increased, there was an increase in both deflection and crack width. Notably, the difference in crack width between NAC and RAC beams grew with higher RCA replacement percentages at the same ρ and concrete grade.

Similarly, Ignjatović et al. (2013) explored the impact of coarse RCA replacement (50 and 100%) and ρ levels (0.28, 1.46 and 2.54%). At higher ρ values (1.46 and 2.54%), inclined cracks become more prevalent in the shear span and caused significant damage, which was more pronounced with higher RCA replacement percentages, in contrast to NAC beams. However, when the ρ values are held constant, there was negligible difference between the ultimate moment and service deflection of coarse RAC and NAC beams. A comparative analysis of various research studies indicated about 13% increase in deflection and 10% reduction in cracking moment (due to 10% lesser tensile strength of concrete) in RAC beams. In addition, Kang et al. (2014) examined the effect of coarse RCA replacement (15, 30 and 50%) and ρ (0.5, 1, 1.5, and 1.8%) and reported a greater number of closely spaced cracks but similar crack width and ultimate flexure strength in RAC beams. The failure pattern in RAC beams was characterized by a significant upward shift of the neutral axis and a higher increase in concrete strain after cracking, attributed to the failure of the aggregate-matrix bond. For satisfactory structural behaviour in concrete, Kang et al. recommended a maximum 30% RCA replacement and proposed a reduction factor based on the RCA replacement percentage to account for interface failures in RAC beams. Knaack and Kurama (2014) reported equivalent flexural strength and failure mode of RAC beams under flexure. However, RAC beams exhibited higher deflection, primarily due to a decrease in stiffness after the initiation of cracking. This behaviour was more pronounced with higher levels of RCA replacement.

Arezoumandi et al. (2015) examined flexure behaviour of RC beams composed entirely of 100% coarse RCA while varying ρ at 0.47 and 0.64%. Their findings also revealed significantly lower cracking moment (about 20 and 10% for different ρ values), equivalent yield moment and either similar or slightly higher ultimate flexural strength as compared to conventional concrete beams. Effect of coarse RCA replacement was reflected in the formation of a greater number of closely spaced cracks leading to reduced modulus of elasticity, moment of inertia and increased ultimate deflection. Both ACI-318 and EC-2 provisions were found to conservatively estimate flexure capacity of RAC beams.

Seara-Paz et al. (2018) delved into the influence of the water-cement (w/c) ratio at 0.5 and 0.65, as well as coarse RCA replacement percentages at 20, 50, and 100% on flexure behaviour of RC beams. Their observation showed a lower cracking moment (13–14% of the ultimate moment) in 100% coarse RAC attributed to a decrease in split tensile strength. Horizontal splitting cracks emerged at higher ρ indicating bond failure between the steel–concrete in RAC beams. Similar crack spacing but higher crack width was noticed in RAC beams emphasizing the need to account for the coarse RCA replacement percentage in standard predictions of crack width. Importantly, flexural strength and service load deflections of beams were not significantly affected by coarse RCA in beams.

Choi et al. (2012) observed that beams composed entirely of 100% coarse and fine RCA exhibited equivalent flexural strength with only 2% variation in strength, deflection as well as in the failure behaviour.

Mix design method was considered as one of the influential parameters besides RCA replacement and ρ by Fathifazl et al. (2009a). They adopted the equivalent mortar volume (EMV) mix design method and analysed RCA replacement percentage at 63.5 and 74.3% along with ρ values at 0.49, 1.99 and 3.25%. Their findings revealed that RAC beams exhibited slightly higher (about 8%) ultimate moment (at average ρ) as compared to NAC beams. However, cracking moment was lower due to the adhered mortar on RCA causing weaker ITZ. Furthermore, the study observed higher ductility and smaller crack spacing in RAC beams. Notably, when applying the ACI (American Concrete Institute) standard equation, there was an overestimation of cracking moments and an underestimation of theoretical deflections. As a solution, the researchers recommended using split tensile strength instead of the modulus of rupture for more accurate predictions of cracking moments.

Previous studies have investigated the reliability of reinforced concrete flexural members, shedding light on the variability introduced using coarse recycled aggregates in concrete. Xiao et al. (2016) reported larger coefficient of variation (COV) of RAC due to the higher variability associated with multiple sources of recycled materials. They recommended increasing both the minimum and maximum tension reinforcement ratio to achieve a similar level of reliability as NAC. Specifically, they highlighted that at higher tension reinforcement ratio, larger COV of recycled materials have greater impact on the flexural reliability of beams. Breccolotti and Materazzi (2013) investigated into the structural reliability of bonding between steel rebar and concrete and observed that the normalized bond strength was similar for both RAC and NAC. They suggested that the limit state design methods, as per EC-2 and model code (MC-2010) could adequately account for the slight bond variability observed in RAC. Similar safety coefficient as that of NAC, can ensure safety for the anchorage and lap splice between steel rebars and RAC. The reliability of Eurocode 2 and ACI provisions was evaluated, and a higher bias factor was obtained for the cracking moment by Pacheco et al. (2019). Notably, the inclusion of Recycled Concrete Aggregate (RCA) had no significant effect on the partial safety factor of RC beams under flexural loading.

8.2 Behaviour of FARAC Beams Under Flexure

The aforementioned experimental studies did not include the incorporation of other supplementary cementitious materials (SCMs) in RAC. To address this gap, Sunayana and Barai (2018) undertook an effort to investigate the flexural behaviour of fly ash incorporated 100% RAC. Additionally, they assessed the serviceability reliability of RAC beams, aiming to account for variabilities associated with predicting the deflection of these beams.

Based on the existing literatures as discussed above, the key parameters under investigation can be identified as follows: (1) longitudinal reinforcement percentage (2) compression reinforcement (3) type of concrete and its material properties and (4) mix proportioning method. The experimental results were analysed and interpreted in terms of (i) load–displacement (ii) failure mode and (iii) moment–curvature relation. Two similar beams were cast for each variation pf parameters. NAC beams without fly ash replacement acted as control specimens. The identifier of beam specimens is as follows: ab-cd, where a: concrete type (NAC, RFA20 and RFA30); b: longitudinal reinforcement percentage (0.42 or 0.75or 1.32); c: indicate whether shear span is provided with stirrups or not ('S' for stirrups present and 'A' denotes absent) and d: serial number (1 or 2). Serial numbers refer to the beams which were similar in terms of their type of concrete mix and reinforcement ratio. In compliance with the Indian standard IS-456:2000, beam specimens of rectangular cross-section having width 200 mm and depth 300 mm were prepared. Length of the specimens were considered as 2.4 m. The ratio between the shear span to the effective depth of the beams (a/d) was taken as 2.62. Tension reinforcement percentages were selected to represent under-reinforced, balanced, and over-reinforced modes of failure. Spacing between the stirrups in the shear span was fixed at 100 mm to avoid the failure of the beams in shear. As the beams were intended to be singly reinforced, the top reinforcement was exclusively utilized to hold the stirrups in place. Reinforcement details of the RC beam are presented in Fig. 8.1. The beams were given a simple support and four-point bending test was performed using a 300-ton universal testing machine. Load was applied at a uniform displacement rate of 1 mm/min. The strain developed in the steel during loading was measured with the help of two strain gauges fixed to the tensile reinforcement at the midspan. A gauge length of 10 mm and 30 mm were used respectively. The vertical deflection was measured at mid-span using linear-variable differential transformer (LVDT).

8.2.1 Load-Carrying Capacity

Table 8.1 provides the experimental findings of the flexural behaviour of beams. The moment value corresponding to the state where there is a change in the initial stiffness of the load–deflection curve is considered as the cracking moment (M_{cr}), indicating initiation of the first crack. The yield moment (M_y) is the point at which deflection

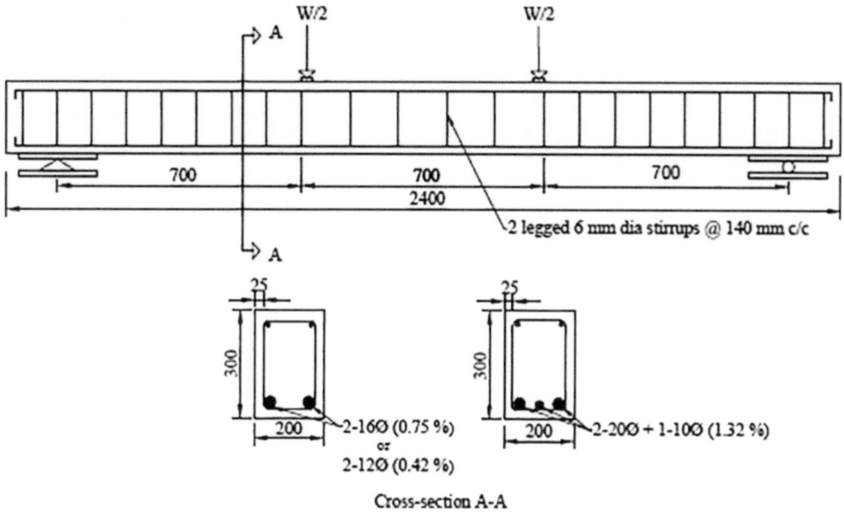

Fig. 8.1 Reinforcement and cross-section of flexure test beams. *Note* Dimensions are shown in mm (Sunayana and Barai 2018)

experiences a sudden increase with a slight increase in the applied stress. The ultimate moment (M_u) is the point on the load–deflection plot where the load reaches its maximum value. Mid-span deflections related to M_{cr}, and M_y are represented as Δ_{cr} and Δ_y, respectively. The maximum value of mid-span deflection is denoted as Δ_u.

8.2.1.1 Effect of Tension Reinforcement

In this section, we will discuss the effect of ρ on the cracking, yielding and ultimate moment of the RC beams. Cracks are initiated when the tensile stress induced by the applied load exceeded the tensile strength of the concrete. According to Indian Standard (IS) provisions, the moment-carrying capacity of an under-reinforced RC beam depends on ρ and the yield strength of the steel, denoted by f_y (as shown in Eq. 8.1, where A_{st} represents the area of tension reinforcement, d stands for the effective depth of the beam's cross-section, and x_u signifies the depth of the neutral axis). It is assumed that the beam does not bear any additional load beyond the point of steel yielding, with the steel primarily contributing to enhanced ductility. On the other hand, the moment of resistance for an over-reinforced RC beam, which is designed to fail under compression, is predominantly influenced by the properties of concrete (Eq. 8.2 where f_{ck} is the characteristic concrete strength). It is to be noted that, at the ultimate failure state, all the beams experience failure due to crushing of concrete.

$$M_u = 0.87 f_y A_{st}(d - 0.416 x_u) \tag{8.1}$$

Table 8.1 Test results on flexure behaviour of beams

Beam ID	Moment (kN-m)			Deflection (mm)		
	M_{cr}	M_y	M_u	Δ_{cr}	Δ_y	Δ_u
NAC-0.42S1	18.67	39.52	50.39	0.56	5.04	32.2
NAC-0.42S2	17.1	40.68	49.57	1.1	7.21	30
NACIS-0.42S1	18.32	38.94	46.83	0.12	4.25	23.96
NACIS-0.42S2	19.78	39.37	46.86	0.09	3.46	29.6
RFA20-0.42S1	13.37	38.28	45.67	0.63	5.93	20.64
RFA20-0.42S2	16.32	40.23	44.5375	0.76	6.06	24.95
RFA30-0.42S1	10.4	34.77	45.74	1.2	5.93	28.8
RFA30-0.42S2	11.33	35.76	46.38	0.9	5.83	29.1
NAC-0.75S1	17.87	70.35	72.38	0.77	8.8	20
NAC-0.75S2	17.54	68.53	70.15	0.72	7.85	17.47
RFA20-0.75S1	19.28	71.74	75.83	0.77	10.05	23.07
RFA20-0.75S2	17.51	68.26	74.55	0.61	8.08	23.46
RFA30-0.75S1	13.52	67.57	77.73	0.77	7.6	25.2
RFA30-0.75S2	14.48	70.05	75.62	0.62	7.29	24.8
NAC-1.32S1	20.15	114.73	118.9	0.11	12.2	17.47
NAC-1.32S2	19.10	113.85	119.16	0.14	9.66	18.28
RFA20-1.32S1	17.5	105.27	107.88	0.24	9.1	16.62
RFA20-1.32S2	18.2	104.74	106.64	0.28	8.4	14.9
RFA30-1.32S1	17	96	101.63	0.15	9.73	15.71
RFA30-1.32S2	16.16	100.71	99.96	0.1	10	19.0

$$M_u = 0.36 f_{ck} b x_u (d - 0.416 x_u) \tag{8.2}$$

Figure 8.2 demonstrates the impact of ρ on the cracking, yielding and ultimate moment of NAC, RFA20 and RFA30 beams. A slightly higher cracking moment was observed at higher ρ values. In the case of NAC and RFA20 beam specimens, there was an insignificant increase (<10%) in M_{cr} with increasing ρ. However, in RFA30 beams, the effect of ρ was more pronounced. For instance, RFA30-0.42A1 specimens showed a lower M_{cr} compared to RFA30-0.75A1 and RFA30-1.32A1 beams. This can be explained by the fact that RFA30 has a lower flexural tensile strength, which leads to a higher induced concrete tensile strain at lower steel strain values (ρ). The effect of ρ was significant to increase M_y and M_u irrespective of the concrete types as indicated in the Table 8.1.

Displacement ductility (μ), defined as the ratio of maximum displacement (Δ_u) to yield displacement (Δ_y), showed a consistent trend of lower ultimate displacement at higher ρ values, regardless of the concrete types. At lower ρ, NAC beams exhibited higher ductility. However, at higher ρ, comparable ductility was observed for RFA20

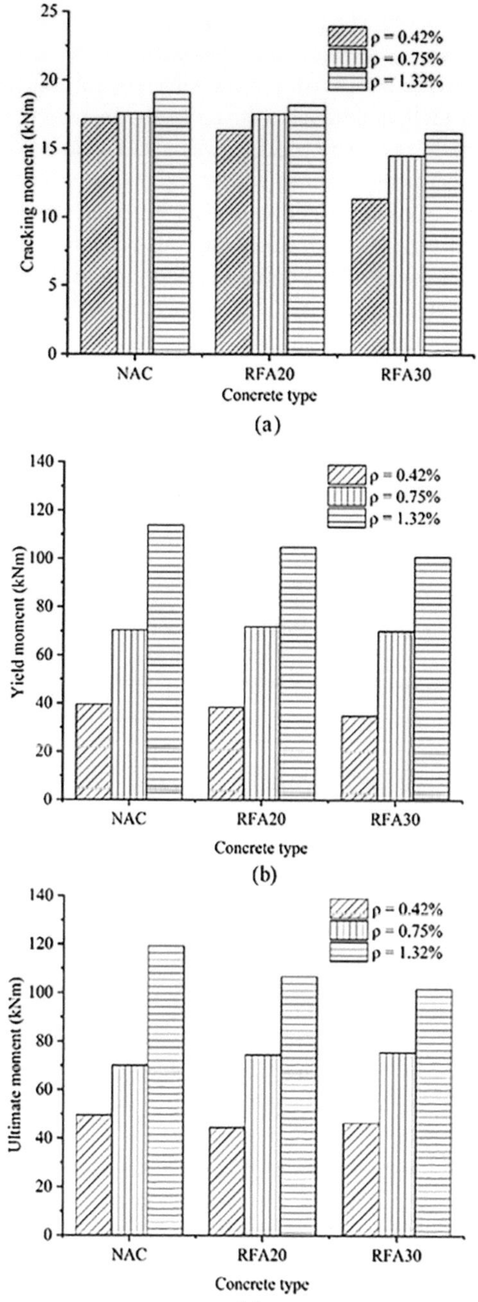

Fig. 8.2 Effect of ρ on **a** cracking **b** yielding and **c** ultimate moment of RC beams (Sunayana and Barai 2018)

and RFA30 beam specimens as well as NAC beams. This trend is further supported by the numerical study discussed in the next section. The differences in ductility observed across the different types of concrete mixes can be related to changes in their individual ultimate loads. Increased peak loads led to greater ductility in the beams. The bond between RAC and deformed steel was stronger in the beam specimens designed with higher ρ (using larger diameter reinforcement bars) owing to improved frictional resistance. This effect of enhanced bond in both plain and deformed bars with RAC was also observed in a pull-out test conducted by Xiao and Falkner (2007). Similarly, According to Corinaldesi and Moriconi (2009), the bond strength in recycled aggregate concrete (RAC) was shown to be around 6% higher without fly ash and 15% higher with fly ash. As mentioned earlier, the triple mixing method forms a layer of fly ash on the aggregate surface and modifies the interface by filling microcracks. Thus, the flexural ductility in FARAC can be equal to or greater than NAC when ρ is increased.

8.2.1.2 Effect of Concrete Type

Figure 8.3 illustrates the load–deflection curves of NAC, RFA20, and RFA30 beam specimens underlining effect of concrete types. All the specimens showed near-linear behaviour up to initiation of crack (elastic region); following which, stiffness of the specimens degraded gradually.

The load–deflection behaviour exhibited a linear increase in deflection with load up to the point of steel yielding, followed by a rapid increase in deflection with smaller variation of load (indicating non-linear behaviour) until the point of failure is reached. At lower ρ (0.42%), there was a marginal decrease (<10%) in M_{cr} in RFA20 beams but a significant (30%) decrease in RFA30 beams as depicted in Fig. 8.2. This can be attributed to the variation in flexural tensile strength (tensile strength values are given in the Chap. 6). Similarly, RFA30 beams showed a reduction of 17% and 15.4% in M_{cr} (compared to NAC) respectively at 0.75% and 1.32% tension reinforcement. However, the differences in yield moment and ultimate moment were relatively insignificant (<10%) among different concrete types, as these values are predominantly determined by the steel reinforcement in under-reinforced beams. In case of higher ρ (1.32%), the compression concrete strain attained a value of 0.003 earlier in FARAC due to a higher rate of increase relative to NAC. Consequently, the ultimate moment carrying capacity was lower by about 10% and 15% in RFA20 and RFA30 beams respectively. To achieve a ductile failure mode, it is recommended to provide adequate compression reinforcement at higher ρ values. Higher deflection was observed at a specific load level up to the yield point in the case of FARAC beams. The occurrence of closely spaced cracks in RAC beams can be related to its lower tensile strength, as depicted in the crack propagation figures discussed in the subsequent section. Consequently, the beam specimens displayed a decrease in their overall stiffness.

At higher ρ, NAC beam specimens exhibited greater ultimate moment and lower service load deflection. The difference in the ultimate moment can be related to the

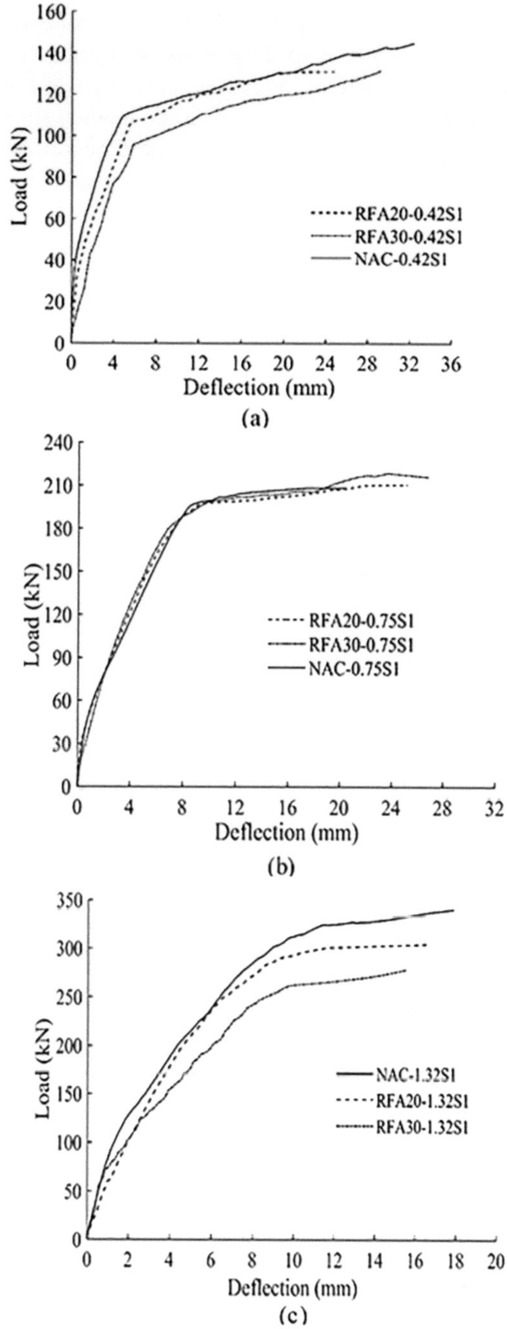

Fig. 8.3 Load deflection curves for beams designed with ρ % **a** 0.42 **b** 0.75 **c** 1.32

variation in the compressive strength of concrete for a specific mix. For FARAC specimens, the mid-span deflection was generally higher (except at the balanced ρ) compared to NAC, although the disparity was not substantial. This effect could be attributed to the enhancement in the modulus of elasticity of FARAC as discussed in Chap. 6. Flexural ductility was mostly equivalent in FARAC and NAC indicating an enhancement in interfacial bond properties in the earlier due to the triple mixing and PPM mix design technique.

8.2.2 Crack Propagation and Failure Pattern

The crack pattern at failure in beams with 0.42% tension reinforcement is depicted in Fig. 8.4. Flexure cracks started forming in the mid-span at a slightly lower load (50 kN) in FARAC specimens compared to NAC (60 kN). With further increase in load, flexure-shear cracks initiated in the shear span. However, the width of flexure cracks increased, followed by the yielding of steel and eventual failure of the specimens. Higher post-cracking stiffness degradation observed in RFA20-0.42S1 and RFA30-0.42S1 in relation to NAC-0.42S1 (as shown in Fig. 8.3) was consistent with the formation of a greater number of closely spaced cracks in the FARAC beams. Crack spacing ranged from 9.46 cm to 17.85 cm (tending towards the higher side) in NAC-0.42S1. In RFA20-0.42S1 and RFA30-0.42S1, crack spacing measured were within the range of 9.2 cm to 15.24 cm and 8.64 cm to 24 cm respectively. Furthermore, it is worth noting that the effect of drying shrinkage might be more pronounced in RAC (Tavakoli and Saroushian 1996). Crack spacing might get decreased and crack count increased as a result of higher shrinkage. Previous studies on natural aggregate concrete found that shrinkage had an impact on the behaviour of cracked concrete (Kaklauskas et al. 2009; Sturm et al. 2017). Additionally, lower tensile strength of FARAC also played an important role in shaping the failure pattern.

As the ρ value increased to 0.75%, a greater number of flexure-shear cracks started developing in the shear span, regardless of the concrete types (as seen in Fig. 8.5). Crack spacing ranged from 13.49 to 24.67 cm in NAC-0.75S1, 10.75 to 16.98 cm in RFA20-0.75S1 and 5.26 to 19.94 cm in RFA30-0.75S1 specimens. Similarly, for RC beams at a higher ρ of 1.32%, failure was marked with crack spacing of 8.98–23.71 cm in NAC-1.32S1, 5.04–16.34 cm in RFA20-1.32S1 and 5.61–15.4 cm in RFA30-1.32S1 (as shown in Fig. 8.6). With increasing ρ, the failure modes transitioned from ductile to brittle. Crushing of concrete occured as the inclined fracture propagated to the compression zone. In addition, a small number of splitting cracks, as well as secondary cracks, were observed along the longitudinal reinforcement at the failure state. These horizontal cracks indicated a failure of bond between concrete and steel. In NAC-1.32S1, although the failure mode leaned towards being brittle, the diagonal cracks did not propagate entirely up to the compression zone unlike in the corresponding FARAC beam specimens. The aggregate interlocking was poor in RAC owing to the presence of adhered mortar content on RCA. As a result, RFA20 and RFA30 beams failed with the formation of prominent brittle cracks.

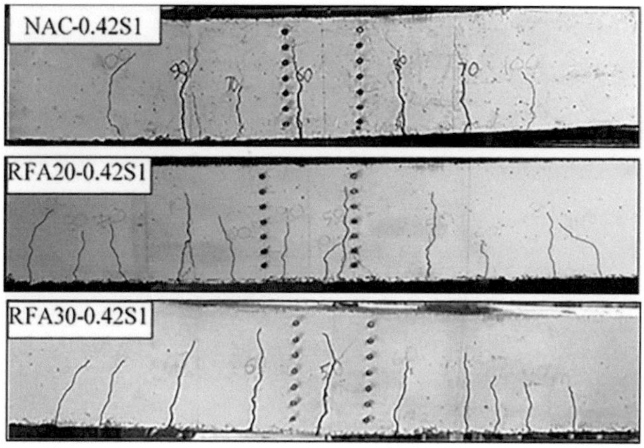

Fig. 8.4 Crack patterns of beams designed with 0.42% ρ (Sunayana and Barai 2018)

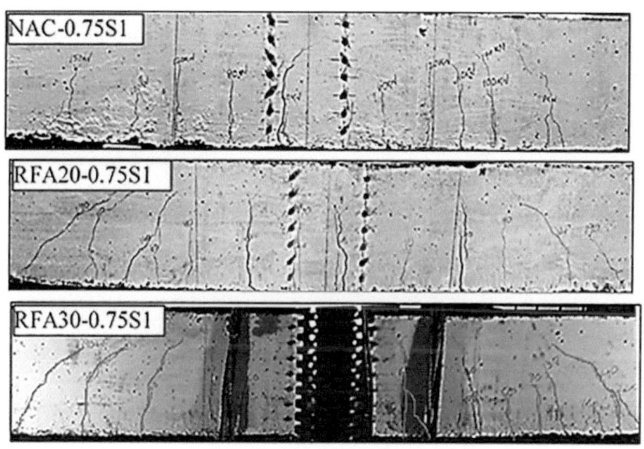

Fig. 8.5 Crack patterns of beams designed with 0.75% ρ

8.2.3 Tension Stiffening of FARAC Beams: A Numerical Investigation

The experimental study on the flexure behaviour of FARAC beams, as discussed in the previous section, revealed slightly higher deflections. It can be noted that earlier studies in this domain reported significantly larger deformations in RAC beams as compared to corresponding NAC beams. Understanding the actual stress–strain relation of cracked tensile concrete and the tension stiffening effect is crucial for predicting deformations accurately. Tension-stiffening can be defined as the tensile stress carrying capacity of the concrete present between two cracks. It is important to highlight that the results of tension tests conducted on plane concrete specimens may

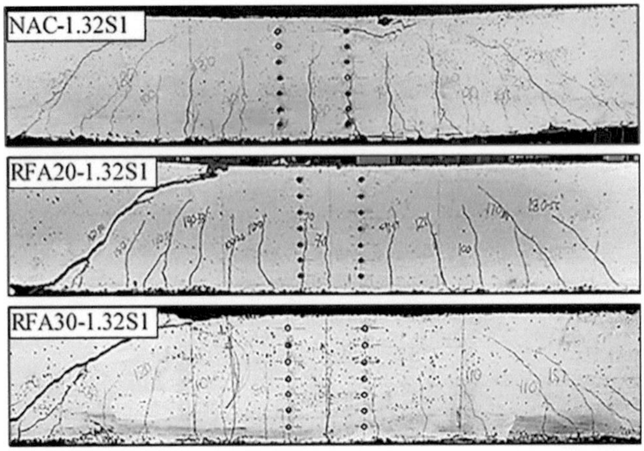

Fig. 8.6 Crack patterns of beams designed with 1.32% ρ (Sunayana and Barai 2018)

not be directly applicable to the bending of beams due to the influence of the concrete-reinforcement bond. Therefore, it becomes necessary to evaluate the actual constitutive relations of cracked concrete. Sunayana and Barai (2018) conducted an inverse analysis using experimentally obtained moment–curvature data to understand the tension stiffening effect in FARAC beams.

8.2.3.1 Methodology

The methodology for assessing the behaviour of cracked tensile concrete can be summarized as follows:

- Moment–curvature relations were computed from experimental load and strain data by applying Euler–Bernoulli beam theory.
- Suitable stress–strain relations were assumed for concrete in compression and steel reinforcement in tension.
- Numerical iteration using inverse analysis was performed with appropriate convergence criteria.

The method employed the smeared crack approach using layered fibre elements. This study assumed the same stress–strain relationship for all fibres within the tensile concrete zone, a linear strain distribution along the depth of cross-section, and a perfect bond between steel and concrete. If we investigate the moment–curvature curves plotted in Fig. 8.7, we can observe the transition of RC beams from un-cracked to cracked phase, marked by a change in initial stiffness. Interestingly, slightly higher curvature was observed in RFA30 beam specimens, regardless of tension reinforcement percentages. Additionally, higher stiffness was reported at larger ρ for the beams, irrespective of the concrete types.

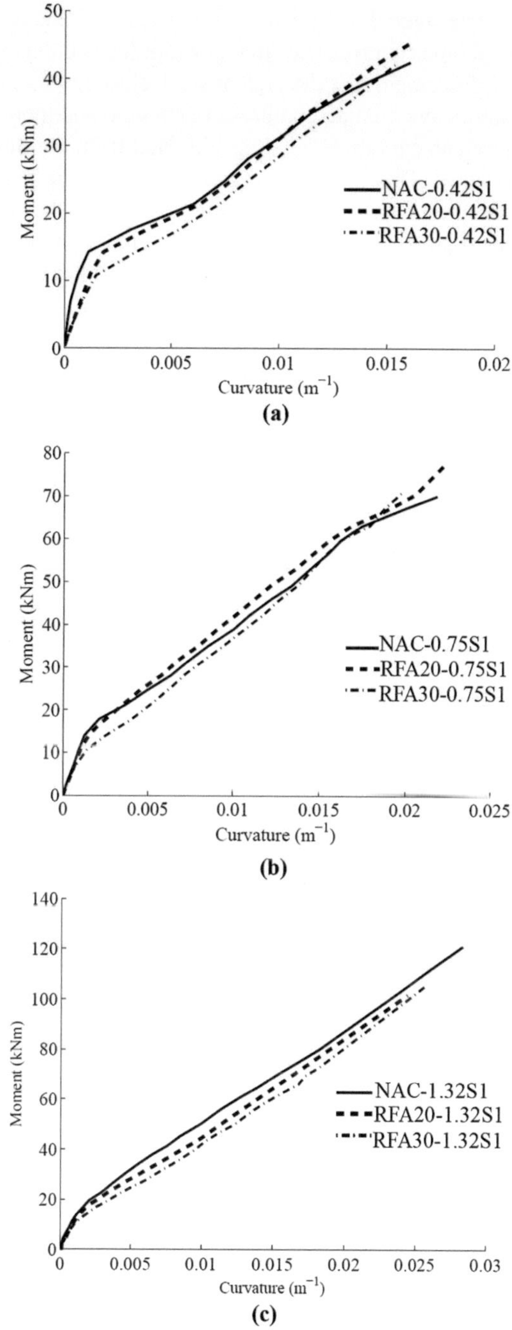

Fig. 8.7 Experimental moment–curvature curves of beams with ρ **a** 0.42% **b** 0.75% **c** 1.32% (Sunayana and Barai 2018)

The layered section model proposed by Kaklauskas (2004) is based on factors such as geometric properties that is width (b) and depth (d) of the beam section, tension reinforcement like the area of steel, elasticity modulus and yield stress. Other input parameters include compressive strength and modulus of elasticity of concrete. The moment–curvature relations obtained from the flexure test the specimens are used as input. This study did not consider the shrinkage phenomenon. Inverse analysis was executed numerically using MATLAB. The layered section model partitions the beam cross-section into a pre-determined number of layers based on the depth of beam cross section. The thickness of reinforcement layer was calculated based on the area of steel. The thickness (t_i) of remaining layers was determined by subtracting the thickness of reinforcement layer from the total depth of the section. Initially, the secant modulus (E_{isec}) value was assumed and assigned to the extreme lower layer in tension. The corresponding curvature was obtained using a direct technique (as described by Gribniak 2009) to calculate strain at that layer for the given curvature using Eq. 8.3.

$$\in = \frac{M}{EI}(y - y_c) \tag{8.3}$$

$$y_c = \frac{\sum_{i=1}^{n} b y_i t_i}{\sum_{i=1}^{n} b t_i} \tag{8.4}$$

$$EI = \sum_{i=1}^{n} \left(bt_i^3 / 12 + bt_i\left[y - y_c^2\right]\right) E_{isec} \tag{8.5}$$

n denotes total number of layers.

A total of 100 layers were taken into consideration for the analysis. The inverse technique combined with the direct method described above, was employed to find out the stress–strain relationship for cracked tensile concrete. For the initial uncracked portion, the experimental curvature was compared with the curvature obtained from the elastic analysis. After cracking occurred, the curvature calculated through the aforementioned process was verified with the experimental curvature within a specified tolerance. At each step, the corresponding $E_{i\,sec}$ was determined for deriving the tensile constitutive relation. An iteration technique based on the Newton–Raphson or bisection method was applied to refine the initial guess of $E_{i\,sec}$ until the tolerance criteria was satisfied. The state of compressive stress in concrete and tensile stress in reinforcement was calculated using the respective constitutive relations by interpolation. Tensile stress corresponding to the strain was computed for each layer using the corrected secant modulus. It is important to note that, all the inner layers in tension zone experienced a state of stress that had already been encountered by the extreme tension fibre. Hence that would be available in the history data of extreme tension layer. Average stress–strain relation of cracked tensile concrete was obtained by solving equilibrium of force and moment for all the layers. During this process, it was observed that the tensile constitutive curves exhibited significant oscillations due to inevitable measurement inaccuracies in the initial experimental data sets. To address this, the curves were smoothened numerically using the exponential moving average technique. The modified-running average method based on Hardy's formula (Pollard 1979) was also used for further smoothening out the oscillations and obtain a reliable constitutive relation for cracked tensile concrete. The constitutive relationship for compressive concrete (NAC) was adopted from Popovics (1973) as shown in Eqs. 8.6–8.8.

$$f = f_c \frac{\in}{\in_0} \frac{n}{n - 1 + \left(\frac{\varepsilon}{\epsilon_0}\right)^n} \tag{8.6}$$

where f_c is the cylinder compressive strength, $\in_0$ is the strain corresponding to peak stress (f_c) which is given

$$\in_0 = 2.7 \times 10^{-4} \times \sqrt[4]{f_0} \tag{8.7}$$

n is the factor signifying slope of ascending branch and is given by

$$n = 0.4 \times 10^{-3} f_0 + 1 \tag{8.8}$$

For RAC, the stress–strain relation for compressive concrete up to peak load was obtained with modifications of peak strain and slope of ascending branch (Mahgoub et al. 2017). Strain at peak stress was considered as 0.0025. The slope of the stress–strain curve was determined following Eq. 8.9. In the post-peak region, stress–strain curve for compressive concrete was modified using Eqs. 8.10 and 8.11.

$$n = \frac{f_c}{2} + 3 \tag{8.9}$$

$$f = f_c - Z_p(\in - \in_0) \tag{8.10}$$

where Z_p is given as

$$Z_p = 400 f_c + 100 \tag{8.11}$$

8.2.3.2 Evaluation of Tension-Stiffening Based on Concrete Types

Applying the above-mentioned procedure, stress–strain curves of cracked tensile concrete were obtained. Figure 8.8 depicts greater tension stiffening for concrete with higher tensile strength (NAC) regardless of reinforcement percentage. Conversely, the lowest tension stiffening was reported in RFA30 specimens primarily due to their lower cracking moment; which subsequently led to higher deflection.

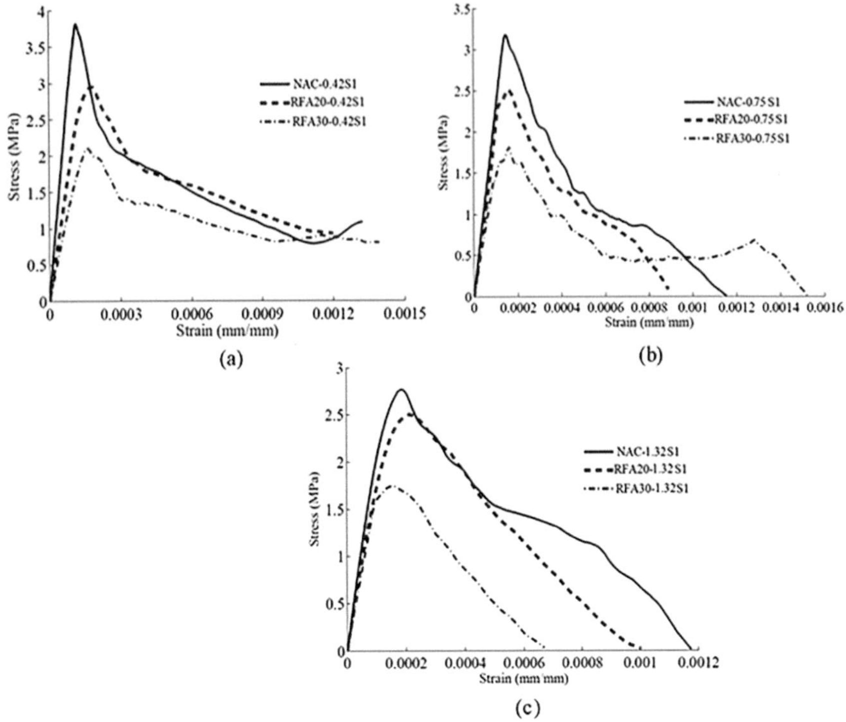

Fig. 8.8 Tensile concrete stress–strain relation for FARAC and NAC

The reasons for these observations can be attributed to factors such as the bond between steel and concrete, and the lower tensile strength of RAC containing 30% fly ash. A similar effect of tensile strength on tension stiffening of concrete was also reported in NAC (Wu 2010). It is important to note that, this study did not consider the effect of shrinkage and creep on the tension-stiffening of concrete. However, the use of coarse recycled concrete aggregates and fly ash in concrete can significantly influence shrinkage and creep, potentially contributing to the lower tension stiffening effect in RC beams. Consequently, there is a need for further research to accurately quantify the shrinkage phenomenon in RAC incorporating fly ash.

8.2.3.3 Effect of Reinforcement Percentage on Tension-Stiffening

The effect of ρ % on the tension stiffening of concrete was observed from the derived stress–strain relationship after the initiation of the first crack. It is important to note that limit state design standards for flexure, according to the existing provisions, often overlook the tensile strength of concrete. However, concrete present in between two cracks also carry some level of stress. This stiffening effect was particularly significant in lower reinforcement percentage (Kaklauskas 2004). This observation aligns with the experimentally observed trend of achieving comparable flexural ductility at higher ρ. As the reinforcement percentage increases, the contribution of steel becomes more dominant, leading to a reduction in tension stiffening (Fig. 8.8). It is worth mentioning that the derived stress–strain relation yielded a lower tensile strength compared to that measured from the indirect tensile test of plain concrete (as discussed in Chap. 6). This difference may be influenced by shrinkage, which could contribute to the deviation in tensile strength. Similar effects have been reported previously in the case of NAC (Gribniak et al. 2009).

8.2.4 Reliability Assessment of Serviceability Limit State of Beams

The service-load deflection observed in RAC beam (both with and without fly ash) specimens was higher than that of NAC beam, as reported in prior studies and reaffirmed in the authors' study as discussed in previous sections. This instantaneous deflection is primarily influenced by the modulus of elasticity and the flexure tensile strength of concrete; both of which also contribute to macrocrack formation. In addition to this short-term deflection, time dependent deformations like creep and shrinkage impact actual deflection of a structural member. Furthermore, variabilities in dimensional factors (such as beam depth and clear cover), material properties including compressive and tensile strength of concrete, and environmental conditions (such as humidity) have profound impact on the total deflection of any structural component (Stewart 1996).

The ultimate limit state describes the failure surface signifying the point at which structural failure occurs. In contrast, the serviceability limit state denotes the condition beyond which, a structure no longer fulfils its functional, aesthetic and comfort requirements. Sunayana and Barai (2021) conducted a reliability analysis on the serviceability limit state for deflection in RAC beams (with and without fly ash) as discussed below. The failure criteria were defined as the point where total deflection surpasses the allowable limiting deflection specified by the standard codes. The probability of serviceability failure is shown in Eq. 8.12 (Stewart 1996).

$$P_f = P[G(\Delta_{limit}, \Delta_{total}(t)) < 0] \tag{8.12}$$

where the limit state function $G = \Delta_{limit} - \Delta_{total}(t)$.

To restrict maximum deflection under flexural loads, various standard guidelines recommend specific span-to-effective depth ratio (l/d) requirements for the RC beam sections. Design of beams and l/d ratio adhered to the criteria set by Australian standard (AS-3600) and Indian standard IS-456 are discussed here. The total deflections accounting for both immediate (elastic) and long-term sustained loads including creep and shrinkage, was kept within a limit of span/250.

AS-3600 prescribes a maximum allowable l/d ratio or minimum depth requirement for a beam section, as described in Eq. 8.13. This formulation expresses l/d as a function of effective design load (considering long-term effects), deflection limit and support conditions (simply supported or continuous), providing a more accurate estimation of total deflection.

$$\frac{l}{d} = \left[\frac{k_1(\nu/l)bE_c}{k_2 F_{d,eff}}\right]^{1/3} \tag{8.13}$$

where $k_1 = 0.02 + 2.5\rho$ for $\rho > 0.005$ (in case of rectangular section).

$k_2 = \frac{5}{384}$ for simply supported beam subjected to uniformly distributed load.

$F_{d,eff}$ = effective design load per unit length considering total deflection (including long-term)

$$F_{d,eff} = (1 + k_{cs})g + (\psi_s + k_{cs}\psi_1)q \tag{8.14}$$

$$k_{cs} = \left[2 - 1.2\left(\frac{A_{sc}}{A_{st}}\right)\right] \geq 0.8 \tag{8.15}$$

The factor k_{cs} is applied to short-term deflection under sustained loads to account for the effects of creep and shrinkage, which are time dependant behaviour. Long-term deflection is commonly considered to be two times as that of the instantaneous deflection ($k_{cs} = 2$) due to sustained loads.

ψ_s = short-term live load factor for maximum annual load = 0.7.

ψ_1 = long-term live load factor considering average sustained load for 50 years = 0.4.

To limit the total short-term and long-term deflection, IS-456 provision suggests that l/d ratio to be governed by factors such as type of support conditions (cantilever, simply supported or continuous), span length if span is over 10 m, the percentage of tension or compression reinforcement and stress levels in steel. A basic l/d ratio is recommended primarily based on support conditions for example, simply supported beam can have a l/d ratio of 20. Additional modification factors need to be applied, taking into account the span length and the percentage of reinforcement area.

The deflection model based on Stewart (1996) computes the total deflection of a simply supported beam accounting for the effect of sustained elastic strain, extraordinary instantaneous, and strains due to creep and shrinkage. The sustained and extraordinary elastic strain were assessed using the principle of structural analysis and mentioned in Eqs. 8.16 and 8.17. The sustained loading causes an instantaneous component of strain which was expressed as a function of time (τ); τ represents the time of initial application of sustained load. Short-term extra ordinary loads, lasting less than a day, caused by a normal or emergency crowding, added an additional 25% strain due to creep effect of extraordinary load (Gilbert 1988).

$$\varepsilon_s(\tau) = \frac{\sigma(\tau)}{E_{conc}(\tau)} \tag{8.16}$$

$$\varepsilon_e(t) = \frac{1.25\sigma(t)}{E_{conc}(t)} \tag{8.17}$$

In addition to the elastic loading, the effects of time-dependant phenomenon such as creep and shrinkage were evaluated using the B4 model (Rilem report 2015). Upon exposure of the concrete surface to atmosphere, the relative humidity decreased, leading to drying. According to the B4 model, magnitude and rate of drying shrinkage strain are influenced by the type of cement (represented by specific constants) and age adjusted elastic modulus (Eq. 8.18).

$$\epsilon_{sh}(t, t0) = \varepsilon_{sh\infty}(t0)k_h S(t) \tag{8.18}$$

$\varepsilon_{sh\infty}$ represents shrinkage correction factor, which is a function of temperature and time and also accounts for variation in cement types, aggregate types and modulus of elasticity (E). The term $S(t)$ denotes the drying rate of exposed concrete surfaces. Additionally, k_h is a constant associated with environmental humidity (h) which is less than 100% after the curing period is over. Detail expressions of these parameters, as well as constants related to cement and aggregate types, are sourced from the 2015 Rilem report. Similarly, the long-term creep strain includes an instantaneous component (q_1), basic creep (C_0) due to sustained loading and drying creep (C_d) corresponding to shrinkage.

$$\epsilon_{cr}(t) = J(t, \tau)\sigma \tag{8.19}$$

where

$$J(t, \tau) = q_1 + C_0(t, \tau) + C_d(t, \tau, t_0) \tag{8.20}$$

Additional explanations and data about model parameters and other crucial factors were obtained from the study of Bažant and Kim (1992) and Rilem report 2015. The mid-span deflection of a simply supported beam under the effects of both short-term and long-term load according to Stewart (1996) are mentioned below in Eq. 8.21 (uncracked case) and Eq. 8.22 (cracked beam).

$$\Delta(t) = \frac{5L^2}{48y(t)}[\varepsilon_e(t) + \varepsilon_s(t)] + \frac{5L^2\varepsilon_{cr}(t)}{48y(t)\alpha_c} + \frac{L^2\varepsilon_{sh}(t)}{8d} M(\tau) < M_{cr}(\tau) \tag{8.21}$$

$$\Delta(t) = \frac{5L^2}{48y(t)}[\varepsilon_e(t) + \varepsilon_s(t)] + \frac{5L^2\varepsilon_{cr}(t)}{48d\alpha_c} + \frac{L^2\varepsilon_{sh}(t)}{8d} M(\tau) > M_{cr}(\tau) \tag{8.22}$$

In the above equations, $y =$ neutral axis depth; d = effective depth of the section; $M =$ moment due to applied load; $M_{cr} =$ cracking moment at time τ for different load

Table 8.2 Statistical properties of basic random variables

Random variables	Bias	COV (%)	Sources
Depth of the beam (d)	1.0	$8.4/d_{nom}$	Ellingwood (1978)
Cover (C)	1.0	$11.2/C_{nom}$	Ellingwood (1978)
Curing and workmanship (k_w)	0.88	0.12	Pham (1985)
Humidity (h)	1.0 (mean = 0.56)	0.1	Stewart (1996)
Flexural tensile strength (f_{ct})	1.0 (mean = $0.69\sqrt{f_c}$)	0.2	Mirza et al. (1979)
Creep and shrinkage model error factor			
Ψ_1	1.0 (mean = 1.0)	0.14	Bažant and Kim (1992)
Ψ_2	1.0 (mean = 1.0)	0.25	Bažant and Kim (1992)
Ψ_3	1.0 (mean = 1.0)	0.14	Bažant and Kim (1992)
Compressive strength			
NAC	1.39	8.3	Xiao et al. (2005)
RAC	1.31	9.2	Xiao et al. (2005)

Note C_{nom} nominal concrete cover; d_{nom} nominal effective depth

cases; α_c = a parameter adopted for tensile reinforcement may be 1.4 or 3 (Gilbert 1988). The deflection of a simply supported beam under uniformly distributed load was evaluated while accounting for variability of influencing parameters, which are listed in Table 8.2. All material parameters were assumed to have a normal distribution. The load distribution parameters are detailed in Table 8.3. A stochastic load distribution was considered over a period of 8.5 years based on the requirements of CEB-FIP, which suggests that structural members should withstand construction loads, environmental factors (e.g. shrinkage), and extreme accidental load. The load cases were selected following Stewart (1996), which was based on the work by Ellingwood and Culver (1977). The loading sequence was as follows:

- After removal of formwork at $\tau 1 = 14$ days, sustained construction load and dead load (including beam self-weight) began to act on the structure and continued for a period of 0.5 years.
- Further, sustained floor live loads were applied on the structure after 0.5 years when the building starts to be occupied.
- Sustained extraordinary loads were applied for a duration of 14 days, and instantaneous extraordinary loads acted along with sustained dead load and sustained floor load. The instantaneous extraordinary loads are caused due to normal and emergency crowding, whereas sustained extraordinary loads accounted for furniture stacking.

8.2.4.1 Calculation of Probability of Failure for Deflection Limit States

The beams were designed for flexure according to AS-3600 and IS-456 while satisfying the respective serviceability (l/d ratio requirements) criteria. Notably, design

Table 8.3 Probability distribution parameters of load

Load	Mean	Variance	Distribution	Sources
Dead load	$1.05 \times g$	COV = 0.1	Lognormal	Pham (1985)
Construction live load	0.3	1.03	Weibull	Karshenas and Aoyub (1994)
Sustained floor live load	0.518	0.133 + (2.2/AI)	Gamma	Choi (1989)
Extraordinary floor live load (EL)	$\alpha_m AI^{\beta m}$	$(\alpha_\sigma AI^{\beta\sigma})^2$	Gamma	Choi (1991)
Maximum instantaneous EL	–	–	$F_{Lemi}(x) = exp\left\{-\sum_{i=1}^{2} v_i T[1 - F_{Lei}(x)]\right\}$, i = 1, 2	Harris et al. (1981)
Maximum sustained EL			$F_{Lems}(x) = exp\{-v_3 T[1 - F_{Le3}(x)]\}$, i = 3	Harris et al. (1981)

Note g = design dead load; AI = influence area = 100 m^2, i = 1 (normal crowding), 2 (emergency crowding), 3 (furniture stacking); v_i = mean rate of occurrence of extraordinary load ($v_1 = 1, v_2 = 0.02; v_3 = 0.2$)

load is not considered when determining the l/d ratio according to IS-456 standard. For sections designed according to the maximum l/d ratio recommended by the code, even with the modification factor, the effective depth often fell short of what was necessary to meet strength requirements. In such cases, to achieve a safe flexure design, the necessary tensile reinforcement at times exceeded the limiting reinforcement ratio for a balanced section, or even surpassed the maximum limit. Therefore, reliability analysis for limit state of deflection under IS-456 was performed for balanced sections by adjusting the l/d ratio according to the strength criteria.

Monte-Carlo simulations were performed with 50 million iterations to generate the basic random materials and load parameters. This analysis considered beams span of 4 meter, 6 meter, 8 meter, 10 meter, 12 meter, and 14 meter. Key material parameters, such as compressive strength, flexure tensile strength, and modulus of elasticity were taken as function of time and derived from Rilem report in 2015. The beam deflection were recorded over 28 days, 90 days, 0.5 year, 1 year, and so on, with intervals of 0.5 years up to a maximum of 8.5 years.

The effect of creep and shrinkage strain on the long-term deflection of RAC beams are substantial. Previous studies reported that RAC experiences higher creep (Domingo et al. 2009) and shrinkage (Fathifazl et al. 2011a; Kou et al. 2007; Kou and Poon 2012; Zhu et al. 2014). Silva et al. (2015) proposed a shrinkage modification factor of 1.8 for concrete with 100% RCA replacement. Lye et al. (2016) recommended creep correction factors based on concrete compressive strength, according to which a factor of 1.41 may be used for 100% RAC with a compressive strength of 30 MPa. The higher creep strain in RAC may be caused due to the presence of adhered mortar, larger porosity, lesser modulus of RCA and use of higher proportion of cement or total mortar content (Agrela et al. 2013; de brito and Saikia 2011). The lower modulus of RCA led to reduced restraint against shrinkage. Since the modulus of elasticity significantly impacts short-term deflection, the EC-2 formula was utilized to determine the concrete modulus. For RAC, EC-2 suggests the modulus to be 0.7 times that of NAC due to increased shrinkage, reduced crack spacing and accelerated stiffness degradation in RAC beams. The serviceability reliability index (ß) was derived based on the probability of failure.

8.2.4.2 Results and Discussion

Figure 8.9 characterizes the simulated probability distributions of total deflection for 4 m and 10 m span beams based on AS-3600 and IS-456. The mean deflection increased over time, with the distribution peak flattening with increasing age, reducing the likelihood of deflections near the mean. The mean deflection predicted by the IS code is higher as compared to that by AS-3600 due to differences in the l/d ratio requirements and effective moment of inertia calculations. Additionally, the IS design led to a higher design load by following a load combination of (1.5 DL + 1.5 LL) in contrast to the load combination of (1.2 DL + 1.5LL) suggested in the AS code.

Figure 8.10 presents the contributions of elastic, creep, and shrinkage components to the total deflection for RAC beams with 4 m and 10 m spans. In the 4 m span RAC beams, elastic deflection initially dominated at 28 days, then decline rapidly at a higher rate up to 500 days and became stabilized afterwards. In contrast, deflection due to drying shrinkage began to rise immediately as curing ended. Creep deflection exhibited an initial decrease followed by an increase over time, ultimately reaching a constant value. Similar trend was also observed in RAC beams of 10 m span. The effect of age on total deflection of NAC and RAC beams is illustrated in Fig. 8.11. Total deflection increased significantly during the initial first year; then at a lower rate thereafter, regardless of the concrete types- a trend also noticed in NAC beams by Stewart (1996). RAC beams notably showed a higher deflection as compared to NAC beams, regardless of span. The elastic component of total deflection was greater in RAC due to its lower modulus of elasticity. Additionally, the higher creep and shrinkage deflection in RAC were attributed to the larger correction factor discussed earlier.

Figure 8.12 shows the l/d ratio required for beams of different span to meet the target reliability index. The l/d ratios calculated for NAC beams according to AS-3600 were also applied to RAC beams. However, the requirements of l/d ratio for beam specimens designed according to IS-456 in NAC was higher than RAC. In this case, the l/d ratio was modified to attain a serviceability reliability index of 1.5. Due to the greater short-term and long-term deflection of RAC beams, lower values of l/d or higher effective depth was suggested as compared to NAC beams. The value of

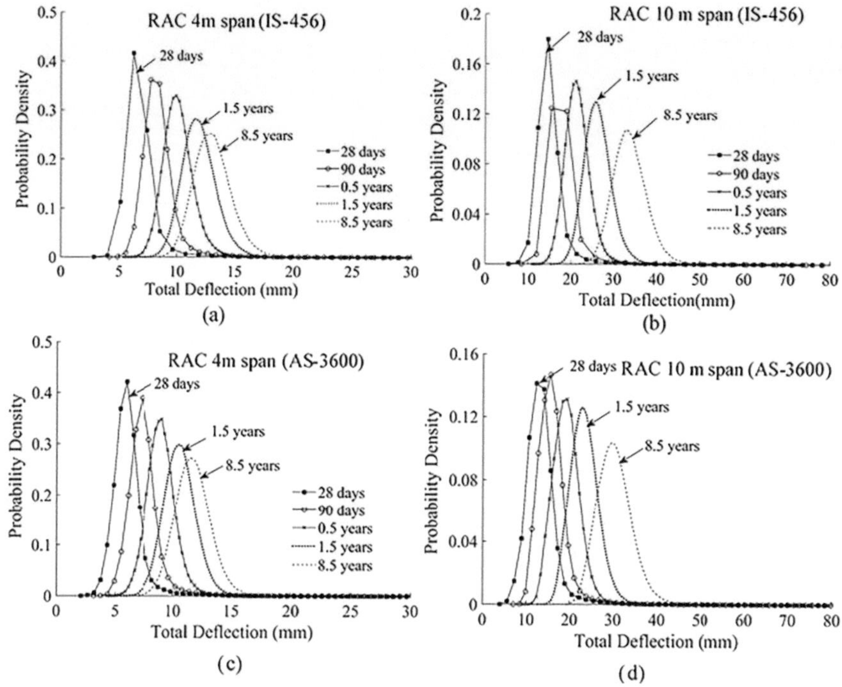

Fig. 8.9 Simulated probability distributions for deflection of RAC beams designed with **a** IS for 4 m span **b** 10 m span **c** AS for 4 m span **d** 10 m span

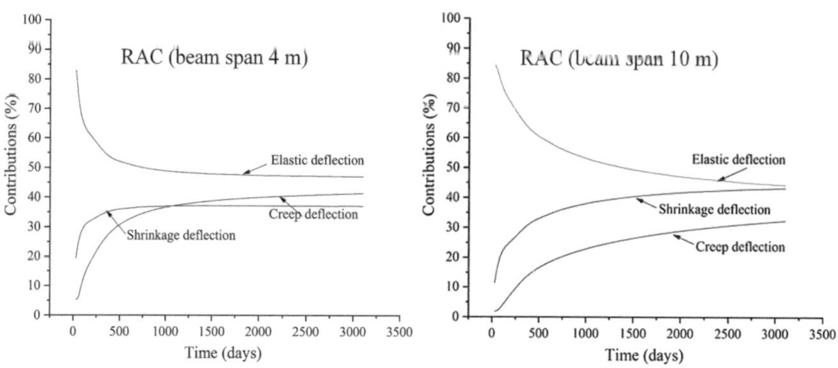

Fig. 8.10 Contributions (%) of elastic, creep, and shrinkage deflection

ß varied over time depending on the beam span as shown in Fig. 8.13. ß increased initially up to 0.5 years after which it decreased with time under the effect of sustained loading. The chosen l/d ratio showed ß in a range of 1.6–1.8 at the end of 8.5 years.

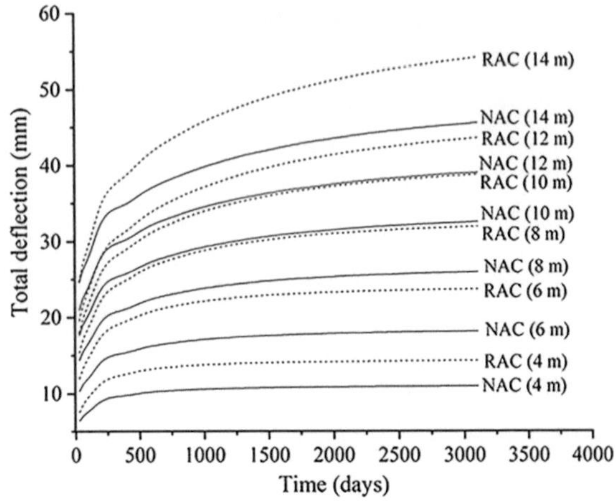

Fig. 8.11 Effect of time on deflection of beams of different span

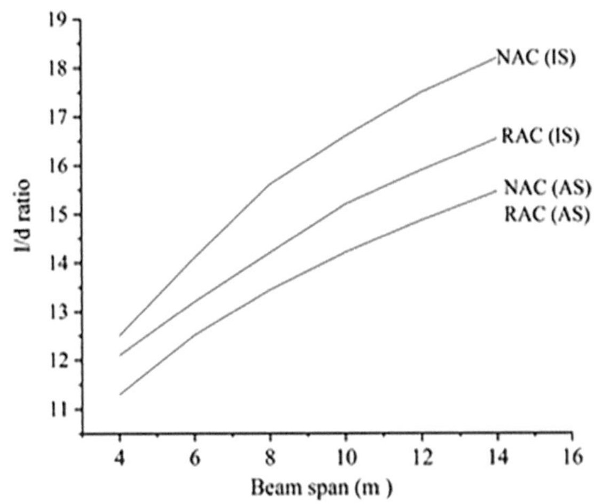

Fig. 8.12 Proposed l/d ratio for different beam span

8.3 Behaviour of Beams Under Shear

This section introduces the topic of shear behaviour in reinforced concrete beams specifically focusing on beams made with coarse recycled concrete aggregate and low-volume fly ash. Reinforced concrete beam designed without transverse reinforcement can be critical in shear due to factors like aggregate interlocking, uncracked concrete and dowel actions. In addition, material properties such as the tensile strength of concrete also affects the shear strength. Tensile strength depends on the type of aggregate and mineral admixture in standard laboratory sample (observed in Chap. 6). Structural behaviour like shear capacity of RC beams may get affected due to FARAC in view of difference in the contributions of shear resisting components.

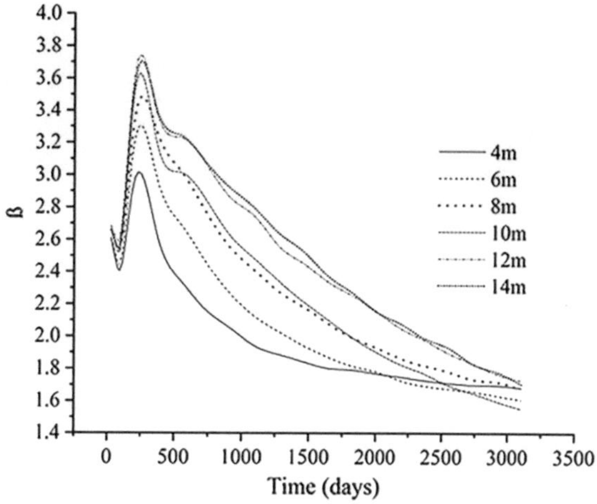

Fig. 8.13 Variation of reliability index with time

In this section, the discussion begins with an exploration of the shear behaviour of RC beams containing coarse recycled concrete aggregates. Then, the study delves into how the inclusion of low-volume fly ash in 100% RCA-based concrete impacts shear resistance in RC beams. Various parameters are considered, including tension reinforcement percentages, fly ash as a cement substitute (at 20 and 30%), and the presence or absence of stirrups in the shear span. The chapter aims to investigate aspects like the initiation of diagonal cracking, failure shear, crack distributions, and critical crack formation in FARAC beams. It also seeks to assess the influence of different parameters, such as aggregate types (100% recycled and natural aggregates), fly ash substitution percentages, and varying amounts of longitudinal reinforcement, on the shear-resisting mechanisms in RC beams. Furthermore, major standard provisions, including ACI-318, EC2, the CEB-FIB model code, and IS-456, which provide guidelines for shear resistance of NAC beams. These provisions are mainly empirical and need to be substantiated before applying them to beams made with RAC and fly ash. How do the shear-resisting mechanisms in RC beams vary due to interaction of parameters like aggregate types (100% recycled and natural aggregates), fly ash substitution percentages, and different amount of longitudinal reinforcement? This question needs to be answered to establish fly ash and 100% RAC as a relevant, safe, and sustainable construction material in practice with ample confidence.

8.3.1 *Shear Behaviour of RAC Beams*

Shear performance of RC beams constituting coarse RCA are briefly discussed, as it's important to understand the role of RCA in shear behaviour. Some of the key parameters that affect shear behaviour include concrete strength, tension reinforcement percentage, shear-span to depth ratio (a/d), and the percentage of RCA replacement.

RC beams designed without stirrups in the shear span are more critical for shear failure. Han et al. (2001) examined the influence of a/d ratio (1.5, 2, 3, and 4) and shear reinforcement percentage (0, 0.089, 0.244, 0.507, and 0.823%). They observed that RAC beams without shear reinforcement (stirrups) exhibited a less inclined critical shear crack with increasing a/d ratio. This means that as the span of the beam relative to its depth increased, the angle of the critical shear crack became less steep. Similar failure mode but a different crack texture (smoother crack) was observed due to the strength variation between aggregate and matrix (variation was smaller in case of coarse RCA as compared to NCA). Poor aggregate interlocking in RAC resulted in lower ductility of beam elements. At low a/d ratio, shear resistance occurred through arch action in both RAC and NAC beams. ACI provision were found to be unconservative in predicting the shear capacity of RAC beams with higher a/d ratio (>3). Application of RCA resulted in 10-20% reduction in shear strength of RC beams due to week aggregate interlocking effect (Maruyama et al. 2004).

Etxeberria et al. (2007) conducted experiments on shear behaviour (failure shear, crack pattern) of RC beams with and without transverse reinforcement. The study found that the shear resistance of RC beams was adversely affected as the percentage of RCA replacement exceeded 25%. Lower diagonal cracking load was observed in RAC beams without stirrups, which indicates that RAC beams without stirrups are more susceptible to shear failure. To mitigate the loss of shear strength in RAC beams, the study suggested a potential measure such as decreasing the effective w/c ratio or increase in cement content. Standard provisions of Eurocode, Canadian standard association (CSA) or modified compression field theory (MCFT) methods overestimated shear capacity of RAC beams when the RCA replacement exceeded 25%. Hence, they recommended a limit of 25% RCA replacement as a safer threshold to ensure both durability and strength requirements. Moreover, the presence of transverse reinforcement (stirrups) in RC beams, even with high RCA replacement percentages (50% and 100%), resulted in equivalent failure shear capacities as that of beams with natural coarse aggregate.

Choi et al. (2010) investigated effect of a/d ratio (1.5, 2.5 and 3.25), ρ (0.53, 0.83, 1.61%) and RCA replacement (30, 50 and 100%) on the shear behaviour of RAC beams. Results indicated reduction in shear capacity as the percentage replacement of coarse RCA in the concrete increased; attributed to the lower tensile strength of coarse RCA as compared to NCA. ACI equation when applied to RAC beams overestimated shear capacity at lower a/d (<2) ratio. They observed that the application of shear provisions from current design standards was limited to concrete mixes with a maximum of 50% coarse RCA replacement. This limitation suggests that the standards did not adequately address the shear behaviour of concrete beams with higher levels of RCA content.

Fathifazl et al. (2011b) applied EMV mix design method, as described earlier, to investigate the shear behaviour of RC beams. They observed higher ductility and about 14% higher ultimate shear strength with an increase in aggregate content. This improvement was attributed to enhanced crack roughness and aggregate interlocking resulting from the use of RCA. Furthermore, at lower a/d ratio, the arch mechanism enhanced shear capacity of RC beams. It was noted that shear predictions based on simplified ACI-318, EC-2 and CSA A23.3-04 were conservative for RAC beams.

Al-Zahraa et al. (2011) reported equivalent shear strength of beams with less than 25% coarse RCA replacement. However, for RAC beams with over 50% RCA replacement, they observed a reduction of approximately 12% in shear capacity. Yun et al. (2011) conducted test on slender beams measuring 6 m in length with an a/d ratio of 5.1 under shear. Surprisingly, they found that the percentage of RCA replacement (30, 60, and 100%) had an insignificant effect on the shear performance of the beams. Applicability of ACI provisions were found conservative for all RAC beams.

Arzeoumandi et al. (2014) explored the effect of ρ by adopting 1.27, 2.03, and 2.71% tension reinforcement in 100% RAC and NAC beams. They reported about 12% reduction in shear strength of RAC beams. This decrease was attributed to several factors, including the weaker ITZ due to adhered mortar content, inferior mechanical properties such as split tensile strength, flexure strength and fracture energy. The relative (with respect to NAC) reduction in fracture energy, split tensile and flexural strength ranged from 7 to 25%, ultimately resulting in the reduction of 11% shear capacity of RAC beams when compared to NAC beams.

Sagher (2016) observed that RAC with 50% replacement of RCA showed lower shear strength as compared to complete replacement due to aggregate interlocking and roughness of crack interface. Moreover, equivalent normalized shear strength was observed in RAC and NAC beams at a/d ratio of 2.5. Katkhuda and Shatarat (2016) treated coarse RCA by soaking in hydrochloric acid to remove adhered mortar and then again treated with sodium penta-silicate to fill the cracks present on RCA surfaces. These treatments resulted in slight increase (about 5%) of shear strength for beams comprising treated aggregates. Shear capacity provisions of ACI, EC and fracture mechanics approach (Gastebled and May 2001; Bazant and Yu 2005) were found to be conservative in case of treated RCA.

Choi and Yun (2017) investigated the shear performance of RAC beams with 30, 60 and 100% replacement and a/d ratio varying as 2, 2.5, 3, 4, and 5. At lower a/d, shear in RAC beams was carried through arch action. At higher a/d, beam action governed the shear carrying mechanism and more splitting cracks developed. Beams with a/d ratio of 2, 2.5, and 4 showed shear compression failure. Diagonal cracking load increased with RCA replacement; but, the failure load decreased with higher shear strain and larger crack width due to variation of aggregate interlocking in beams comprising RAC. At constant ρ, similar shear behaviour in terms of shear strength and failure mode were observed in RAC and NAC beams. At low a/d ratio (less than or equal to 2), the ACI equation conservatively predicted shear strength. Sadati et al. (2016) reported 10% reduction in shear capacity of beams comprising 50% RCA and 50% flyash.

Ignjatović et al. (2017) reported that slender RAC beams, designed without stirrups at higher ρ (4%), exhibited an equivalent shear failure. The direction of principal stresses calculated using the strain data in the strain rosette, revealed a slightly greater angle of critical shear crack (about 44°) in RAC beams as compared to NAC beams

(35°). Smoother crack faces in RAC beams suggested less effective aggregate interlock. Regardless of the analysed codes, a 5–6% reduction in the test-to-predicted ratio of mean shear strength was reported in RAC beams.

Rahal and Alrefaei (2017) experimented on the effect of RCA replacement (10, 20, 35, 50, 75 and 100%) on beams constituted of RAC. They observed a 13–18% reduction in shear strength as compared to NAC. A positive effect of replacing smaller size aggregates was reported. To account for the adverse influence of RCA on shear strength, they reduced the $\sqrt{fc}$ term in the shear capacity equation by 20%.

Wardeh and Ghorbel (2019) observed about 25% less fracture energy in 100% RCA replaced concrete. Under four-point bending they reported shear-compression failure in the RC beams designed with an a/d of 1.5 and diagonal tension failure occurred at an a/d ratio of 3. Microscopic observation of crack pattern revealed denser crack with higher crack width in RAC beams. Shear strength of RAC beams (a/d = 3) reduced by 23% compared to NAC beams with similar concrete compressive strength. This decrease in shear strength was attributed to the lower tensile strength of RAC. Notably, the effect of RCA on structural reliability of RC beams for limit state of shear had not been reported previously.

8.3.2 Shear Behaviour of FARAC Beams

Sunayana and Barai (2020) investigated shear resistance of fly ash incorporated RAC beams (referred to as FARAC), emphasizing critical shear aspects. The study assessed shear contributing mechanisms through the analysis of crack distributions and the formations of critical diagonal crack at failure, highlighting the effects of key parameters such as (i) tension reinforcement ratios (2) fly ash substitution percentage up to 30% in complete recycled coarse aggregate based concrete and (3) effect of stirrups. Assessing the shear design provisions of major standards for fly ash incorporated RAC, along with an existing RAC database, is intended to facilitate safe design practices. Twelve beam specimens were designed, constructed, and tested without including transverse reinforcement in the shear span. To account for inherent heterogeneity in concrete, two duplicate beams were fabricated for each set of parameter change. The beam specimens measured 2400 mm span, with cross-sectional dimension of 200 mm in width and an overall depth of 300 mm. Figure 8.14 provides the longitudinal and cross-sectional detailing of the RC beam specimens. Two variations in tension reinforcement ratio (ρ), such as 0.75 and 1.32% were tested. For the 0.75% reinforcement, two 16 mm diameter bars were used, while for 1.32%, two 20 mm bars and one 10 mm diameter bars were considered. The use of higher longitudinal reinforcement ratio ensured that shear failure occurred prior to the flexural failure which commonly occurred in low reinforced concrete beams. The shear span, defined as the area between the load application point and the support, had a constant shear span-depth ratio (a/d) of 2.62, which is greater than 2.5 to satisfy slender beam condition for all beams. No transverse reinforcement was provided within the shear span to assess how concrete properties influence the shear resistance of RC beams.

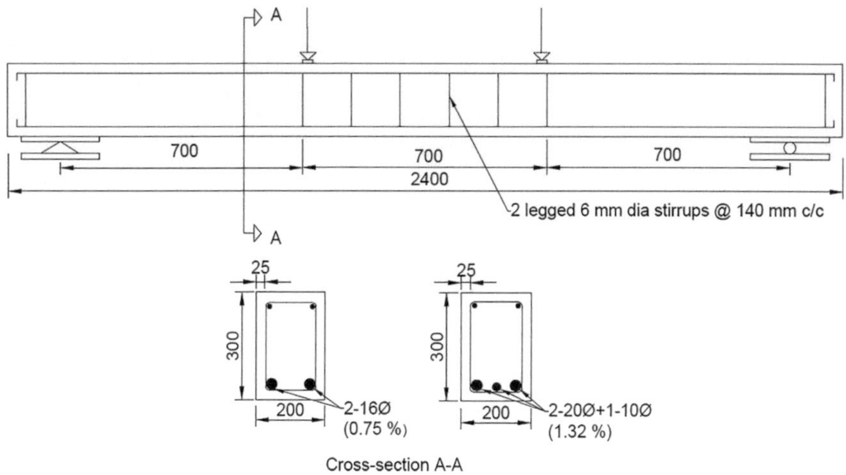

Fig. 8.14 Reinforced concrete beam details

Additionally, 6 mm diameter stirrups were provided at 140 mm c/c interval in the mid-span to prevent localized crushing of concrete beneath the loading point.

8.3.2.1 Load-Carrying Capacity

Figure 8.15 shows the mid-span deflection of RC beams designed with 0.75% longitudinal reinforcement. The region 'A', corresponds to the initiation of flexural cracking marked by an initial stiffness change. At this point, the load capacity in NAC beams was approximately 11–13% greater compared to RFA20 and RFA30 concrete beam specimens. Following the onset of flexural cracking beyond region A, the RFA30-0.75A beam showed a more significant reduction in stiffness attributable to its slightly lower E value compared to NAC. The distribution of numerous closely spaced cracks, as evident in the crack pattern, further explained the reduced stiffness in the RFA30-0.75A beam. Each curve beyond region A followed a linear slope until a sudden drop in load occurred with minimal deformation increase (0–0.23 mm) at point B, marking the onset of critical diagonal crack. This cracking initiated at approximately 93% and 89% of the respective peak loads for NAC and FARAC beams respectively. The RFA20-0.75A and RFA30-0.75A beams demonstrated a shear force approximately 4% lower than that of the NAC-0.75A beam.

The abrupt load drop in the post-peak portion indicated a brittle failure mode of the beams in shear. Following the onset of critical shear crack (section after point 'B') the mid-span deflection was more pronounced for RFA20-0.75A and RFA30-0.75A beams compared to the NAC-0.75A beams. This indicated an increased redistribution of stresses within FARAC beams. The aggregate interlocking reduced and most of the shear resistance was provided by the concrete within the uncracked compression zone alongside the dowel action from the tension reinforcement. Following

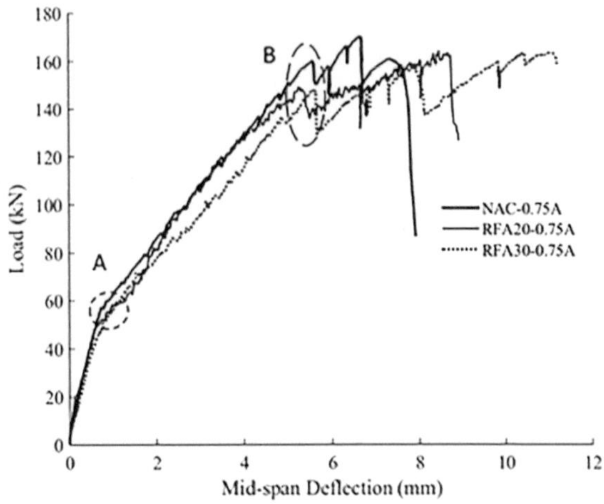

Fig. 8.15 Load–deflection curves of beams with 0.75% tension reinforcement

the formation of diagonal crack, these beams continued to bear additional load and deformation due to the establishment of a stable strut and tie system, or arch action. Typically, the arch action becomes significant in normal concrete beams with an a/d ratio below 2.5 (Tung and Tue 2016). However, this study also observed it even at a higher a/d of 2.62, especially in FARAC beams, as indicated by the load and deformation increase after the drop at point B. The mid-span deflections at the point of critical shear crack formation were recorded as 5.55 mm, 5.25 mm and 5.63 mm for NAC-0.75A, RFA20-0.75A and RFA30-0.75A specimens respectively. Following the development of critical crack, additional deflection was observed in the RC beams amounting to approximately 2.3 mm for NAC-0.75A, 3.6 mm for RFA20-0.75A and 5.5 mm in RFA30-0.75A beams. The corresponding ultimate deflection values were about 7.92 mm, 8.92 mm, and 11.20 mm respectively for these concrete types. The higher deflection in FARAC beams were accompanied by higher critical crack width. The detailed responses of beam specimens under shear are presented in Table 8.4. Due to the inherent heterogeneity in concrete properties, even similar beams displayed slight differences in performance.

Figure 8.16 presents the load–deflection behaviour of beams designed with 1.32% steel reinforcement. The load–deflection curves showed a gentler slope in the FARAC beam specimens. Points A marks the onset of flexure cracking. Points B1 and B2 indicates diagonal cracking in NAC-1.32A and FARAC beams respectively. Notably, the regions B1 and B2 cover a broader zone than region B of Fig. 8.15. The NAC-1.32A beam specimen exhibited a higher load at diagonal crack initiation, this was not easily identifiable on the curves (hence considered as the point at which the diagonal crack was visible for the first time). RFA20-1.32A concrete beam displayed a 7% decrease compared to NAC-1.32A. However, the shear capacity reduced substantially by about 20% in RFA30-1.32A beam. This reduction in failure shear in RFA30-1.32A beam highlights the influence of concrete properties on shear-resisting mechanisms especially at higher tension reinforcement.

Table 8.4 Shear test results (Sunayana and Barai 2020)

Specimen type	Cube compressive strength on beam test day (MPa)	Flexure cracking load V_{fl} (kN)	Shear cracking load V_{sh} (kN)	Shear force at failure V_u (kN)	Shear Strength τ_u (MPa)	Normalized shear strength	Crack width (mm)
Reinforcement percentage-0.75%							
NAC-0.75A	49.08 (1.92)	28.52	79.63	85.13	1.59	0.23	6
NAC-0.75B	48.43 (1.85)	28.50	78.25	86.30	1.61	0.23	6
RFA20-0.75A	47.26 (1.62)	25.08	73.43	82.15	1.54	0.22	8
RFA20-0.75B	46.25 (1.47)	25.10	75.00	84.45	1.58	0.23	8
RFA30-0.75A	45.55 (0.74)	24.65	73.57	81.80	1.53	0.23	9
RFA30-0.75B	44.31 (1.73)	24.60	71.75	80.30	1.50	0.22	9
Reinforcement percentage-1.32%							
NAC-1.32A	52.00 (0.66)	26.65	94.32	109.80	2.05	0.29	4
NAC-1.32B	50.62 (1.17)	26.52	92.47	107.64	2.02	0.28	4
RFA20-1.32A	46.11 (0.95)	21.30	74.12	101.72	1.90	0.28	6
RFA20-1.32B	47.67 (1.25)	21.65	77.48	107.11	2.00	0.29	6
RFA30-1.32A	47.60 (1.7)	20.00	71.00	87.49	1.64	0.24	7
RFA30-1.32B	46.22 (0.21)	19.43	73.23	90.14	1.69	0.25	7

Note The values in the bracket in column 2 indicate standard deviation

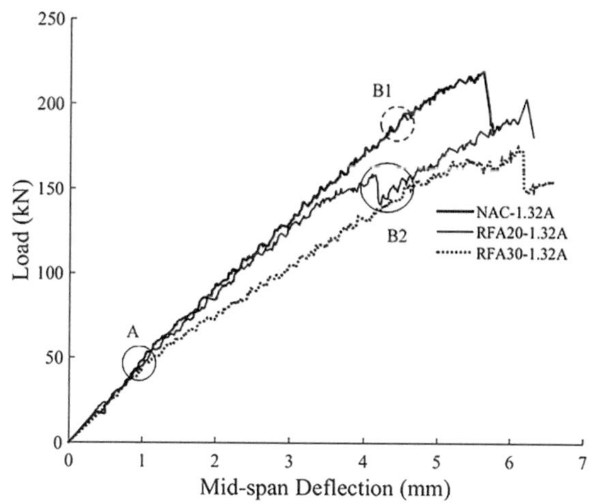

Fig. 8.16 Load deflection curves at 1.32% reinforcement

8.3.2.2 Effect of Tension Reinforcement Percentage

Increasing the tension reinforcement ratio from 0.75 to 1.32% improved the ultimate shear strength of RC beams. Table 8.4 details the experimental results. It shows that shear strength increased by 0.46 MPa in NAC-1.32A compared to that in NAC-0.75A specimens. In contrast, FARAC beams displayed comparatively smaller rise in shear strength for increasing the tension reinforcement. Precisely, the shear strength of RFA20-1.32A and RFA30-1.32A specimens increased by approximately 0.36 MPa and 0.19 MPa, respectively when compared to the corresponding RFA20-0.75A and RFA30-0.75A specimens. This smaller increase in the FARAC shear strength at higher tension reinforcement levels suggests a reduced contribution from the dowel action. Variation in shear capacity of concrete may be attributed to decreased compressive strength, prompting the calculation of normalized shear strength (relative to $f_{ck}^{0.5}$ where f_{ck} is the concrete cube strength on the beam test day). Interestingly, the normalized shear strength values at 0.75% reinforcement were equivalent across concrete types. However, the normalized shear strength of RFA30-1.32A beams was lower than that for NAC-1.32A and RFA20-1.32A beams, possibly indicating that normalizing by $f_{ck}^{0.5}$ may be inadequate (Rahal and Alrefaei 2017) or that shear strength may not be directly expressed in terms of compressive strength under similar tension reinforcement.

8.3.2.3 Effect of Stirrups

In this section, let us discuss the impact of stirrups on the shear behaviour of FARAC and NAC beams by comparing the load–deflection response and crack patterns with and without stirrups. In the beam identifier, the letter 'S' denotes the presence of stirrups within the shear span. Figure 8.17 depicts the load–deflection response of RC beams with and without stirrups. Up to the onset of flexural cracking at mid-span, all specimens exhibited similar behaviour, irrespective of stirrups presence. The addition of stirrups in beam resulted in a well-defined yield point which signifies a ductile failure mode. The transverse reinforcement notably increased load-carrying capacity indicating enhanced shear resistance and ultimate deflection denoting higher ductility of beam specimens. Peak load capacities in RFA20-0.75S, RFA30-0.75S and NAC-0.75S specimens increased by 28.2%, 33.6% and 22.3% respectively as compared to the beams without stirrups. Addition of stirrups also significantly improved ductility with increase of approximately 181% in RFA20-0.75S and 157% in NAC-0.75S beams. However, the ductility gain in RFA30-0.75S beams was relatively smaller (139%) due to the relatively higher ultimate deflection observed in RFA30-0.75A beam.

The presence of stirrups in the case of 1.32% tension reinforcement improved the load carrying capacity to a greater extent (similar or more than 50%). However, the ductility increase due to stirrups was greater for NAC-1.32S beams (209% higher than NAC-1.32A beams) as compared to FARAC beams (163.8% increase in RFA20 beams and 134.8% higher in RFA30 beams).

Fig. 8.17 Load deflection showing effect of stirrups for different concrete

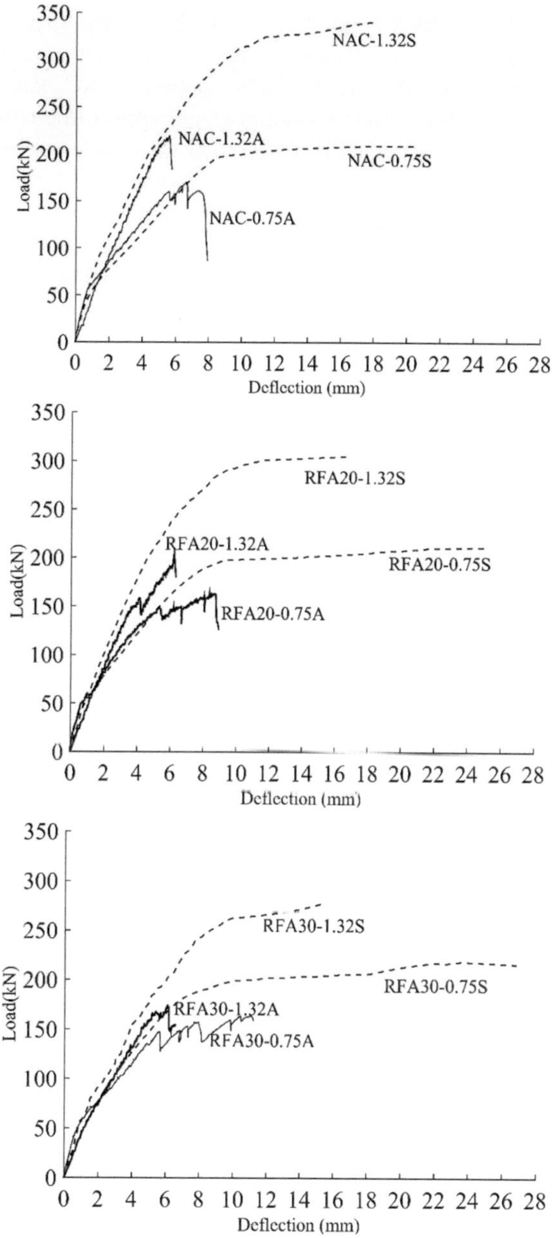

8.3.2.4 Crack Propagation and Failure Pattern

Now, let us examine the failure pattern in RC beams designed without stirrups. Upon loading, discrete flexural cracks appear in the maximum moment zone when the bending tensile stress exceeded the tensile strength of concrete. In the NAC-0.75A specimens, flexural cracks formed at a load of 70 kN, which represented 41% of maximum load capacity. In contrast, crack initiation occurred at 60 kN in RFA20-0.75A and RFA30-0.75A beams, equating to approximately 36% of their peak loads. The early onset of the flexure cracking in FARAC beams is likely due to the reduced tensile strength of concrete. These cracks began internally with micro-cracking but damages became visible later, as indicated by the lower crack initiation load on the corresponding load–deflection curve. Under two-point loading, the bending moment remains constant in the middle 1/3rd of the span, with no shear force present. However, shear force develops in the shear span due to the change in the bending moment, increasing from zero at the support to its maximum at the load application point, causing formation of flexure-shear cracks. Figures 8.18 and 8.19 depict the failed RC beams, showing only the right half portion of the beam due to the symmetry of the specimen. Distribution of cracks at the failure stage is depicted in Figs. 8.20 and 8.21. As shown in Fig. 8.20, the critical crack in NAC-0.75A beams appeared 189.75 mm from the right support, inclined at an angle of 45.7° to the longitudinal axis of the beam, with an average width of approximately 6 mm. The major diagonal crack in RFA20-0.75A beam specimen initiated at 263 mm from the support at an angle of 37° with the axis of the beam. At the failure stage, the crack width at the crack initiation point averaged 8 mm. Both the RFA20-0.75A and RFA30-0.75A beam specimens exhibited sudden and rapid spalling of concrete. The failure of RFA30-0.75A beam occurred with formation of a primary diagonal tension crack that originated 210 mm from the support at an inclination of 31°. This critical crack at the failure was about 9 mm wide. In beams without stirrups, frictional resistance from the protruding aggregate surface helped to resist sliding and limit the opening of the critical crack passing through it. However, in RCA the diagonal crack faces are smoother due to the presence of microcracks in the adhered mortar layer (Han et al. 2001). This reduced frictional resistance in RAC lowers the aggregate interlocking effect. This leads to formation of wider diagonal crack and lower residual stress transfer within FARAC concrete.

The shape and location of the critical shear crack vary as a function of percentage of longitudinal reinforcement. With the increase in main reinforcement percentages, cracks with relatively lower crack width are formed. The inclination of the major diagonal cracks was comparable for NAC and FARAC beams. In the NAC-1.32A beam, the critical crack formed 180 mm from the support at an angle of 29.5° near the crack initiation point. For the RFA20-1.32A specimen, the diagonal crack formed at an angle of 30.5° and 196 mm from the support. In the RFA30-1.32A specimen, the critical crack was observed 222 mm from the support, initiating at an angle of 31.7° relative to the beam longitudinal axis. The crack width measured approximately 4 mm, 6 mm and 7 mm in NAC-1.32A, RFA20-1.32A and RFA30-1.32A specimens respectively at the failure state. As the loading increased, the critical

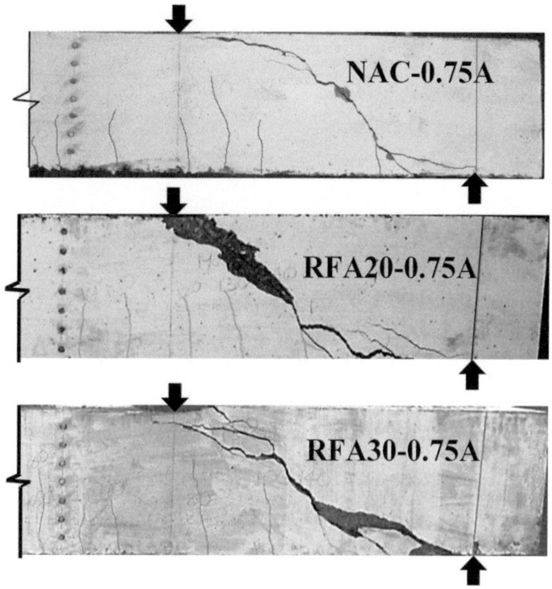

Fig. 8.18 Failure pattern of RC beams designed with 0.75% reinforcement

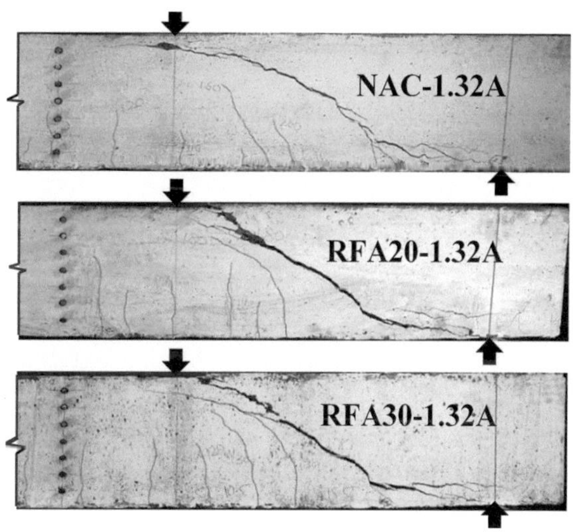

Fig. 8.19 Failure pattern of RC beams designed with 1.32% reinforcement

cracks extended towards the loading point, reducing the compressive concrete area. Simultaneously, horizontal cracks started growing horizontally, mainly in case of FARAC beams, extending towards the support and indicating loss of bond between steel and surrounding concrete. The wider shear cracks in FARAC beams suggested reduced aggregate interlocking in RCA. Beam failure under shear indicated collapse of all the shear resisting mechanisms, regardless of the concrete types. At this point, the shear resistance from uncracked compression concrete was entirely diminished,

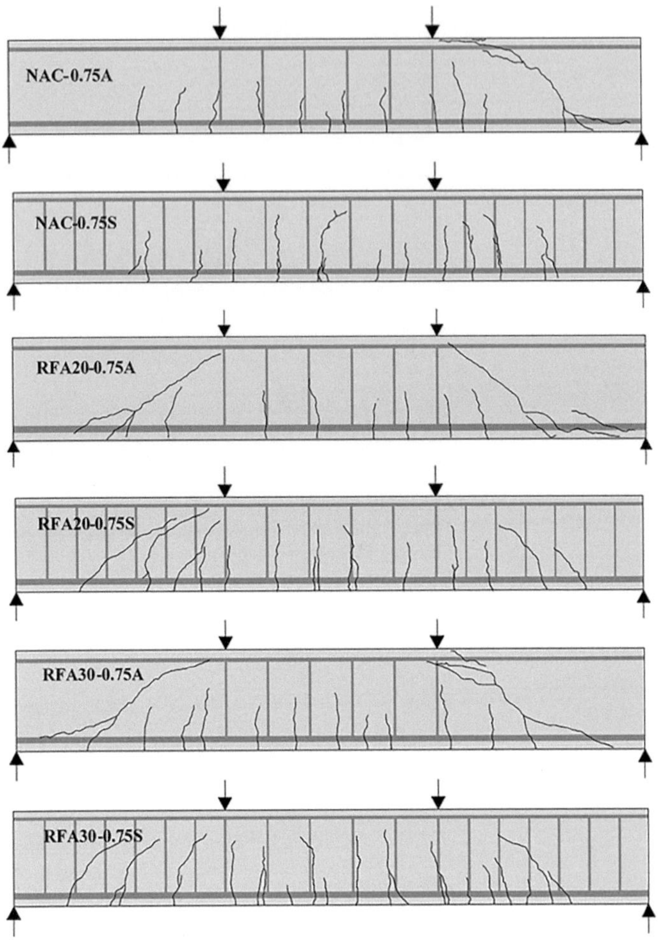

Fig. 8.20 Distribution of cracks in beams with 0.75% ρ with and without stirrups

as evidenced by the critical crack extension up to the loading point in FARAC beams. Additionally, few horizontal secondary cracks appeared in the NAC specimens, within the compression zone near the critical cracks, as shown in Figs. 8.20 and 8.21.

An investigation into the crack formation and distribution in RFA20-1.32A and RFA30-1.32A beams represented in the Fig. 8.20 revealed two key differences compared to NAC-1.32A. In addition to the primary diagonal crack, a flexure shear crack formed nearby, with its upper section closely paralleling the primary crack in RFA20-1.32A and RFA30-1.32A beams. The tortuosity (non-linearity) of the crack path in FARAC beams was lower compared to NAC-1.32A beams. This led to smoother crack propagation in RAC beams attributing to microcracks in the mortar adhered to RCA, which merged to form the major diagonal crack, as seen in the

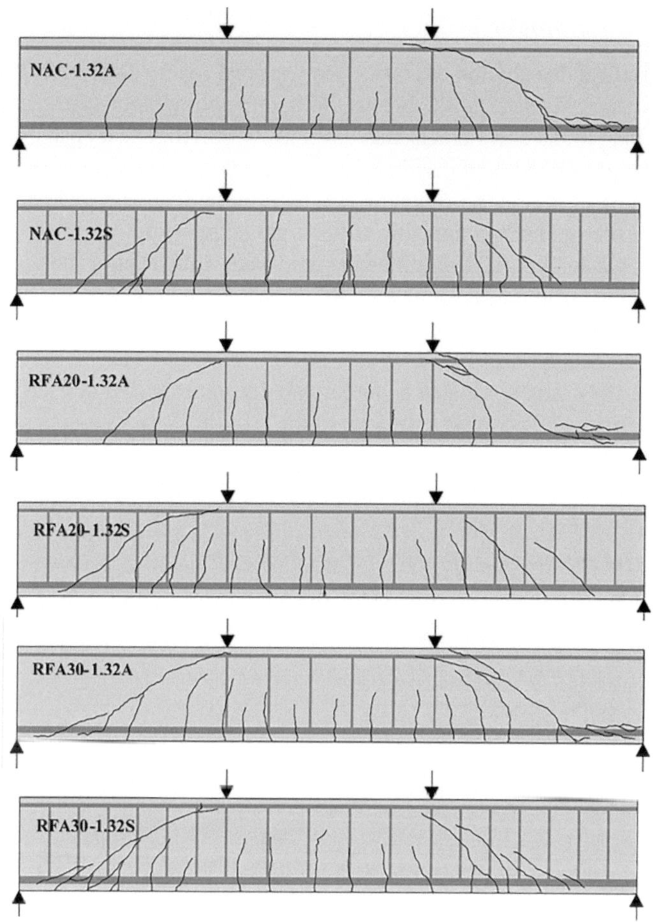

Fig. 8.21 Distribution of cracks in beams with1.32% reinforcement with and without stirrups

overall macro response of concrete. The reduced split tensile strength of RFA20 and RFA30 concrete could influence the shape of critical crack. Principal tensile stress develops in the cracked tension zone within the shear span, especially near the flexural crack tips in this region. The conditions of principal tensile stress exceeding the tensile strength of concrete, marks an area with potential critical crack formation. These flexural crack tips in the highly stressed region eventually merge and propagate towards the loading point, forming inclined cracks. Consequently, in concrete with reduced tensile strength, a larger potential critical crack formation zone emerges, specifically if the zone surpasses the average crack spacing as discussed by Yang (2014) and Tung and Tue (2016).

Analysis of Crack Growth

The initiation and distribution of cracks play a crucial role in determining the failure mode and shear resistance of RC beams, both with and without stirrups. A detailed analysis of the critical shear cracks that develops within the shear span of specimens without stirrups is illustrated in Fig. 8.22. This portion of the beam can be conceptualized as a truss, where the concrete acts as compressive strut and the steel rebar serves as tensile tie, both connecting at the node (support point). The compression strut, illustrated in Fig. 8.22, outlines the load path, transferring the applied force from the loading point to the support.

In NAC-0.75A beam, the critical crack made a steep angle with the beam axis closer to the support. Conversely, in RFA20-0.75A and RFA30-0.75A beams, the diagonal shear crack initiated at a relatively lesser slope and at a larger distance from the support. The position and slope of these critical cracks emphasize two key points:

(1) Steeply inclined cracks are more susceptible to crack edge sliding compared to flatter cracks, which tend to experience more crack opening than sliding (Huber et al. 2016). Consequently, the critical crack growth with a shallower inclination in FARAC beams led to wider crack openings.

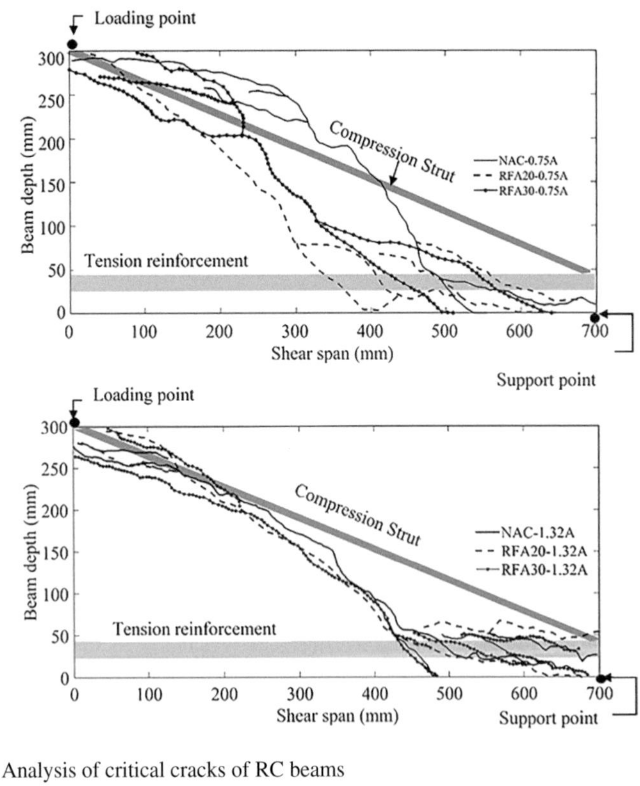

Fig. 8.22 Analysis of critical cracks of RC beams

(2) The activation of dowel action is influenced by the distance of the crack from the support (Campana et al. 2013). Cracks that initiate farther from the support suggest a reduced role of dowel action in the FARAC beams with the same tension reinforcement. Critical cracks that align with the theoretical inclined compression strut indicate a breakdown in the strut-tie shear transfer mechanism. The upper segments of critical cracks in RFA20-0.75A and RFA30-0.75A beams follow the compression strut.

In RC beams with 1.32% tension reinforcement, the critical cracks advanced towards both the loading and support points as the applied loading increased. Numerous horizontal splitting cracks appeared, suggesting bond failure in FARAC beams. The critical cracks extended towards the loading point along the theoretical strut path regardless of the concrete types. Crack locations were largely consistent for NAC-1.32A, RFA20-1.32A, and RFA30-1.32A beams, contrasting with the differences seen in beams with lower tension reinforcement. With a larger bar diameter in 1.32% tension reinforcement, the dowel action is enhanced. However, the crack locations remained relatively consistent across NAC-1.32A, RFA20-1.32A and RFA30-1.32A beams. The overall crack distribution in the beams with and without stirrups is shown in Figs. 8.18 and 8.19. Beams with an equal percentage of transverse reinforcement displayed distinct crack growth patterns based on the concrete types. The failure state of NAC-0.75S beam showed prominent long flexural cracks, while NAC-0.75A beams developed critical shear crack alongside shorter, distributed flexural cracks. The RFA20-0.75A beam showed a major diagonal crack with fewer flexural cracks unlike NAC-0.75A beam. In the RFA20-0.75S specimen, addition of stirrups altered the failure mode to a more ductile one, promoting additional flexural cracks and some flexure-shear cracks. RFA20-0.75S and RFA30-0.75S beams developed flexure-shear cracks with steep slopes intersecting two to three stirrups, unlike NAC-0.75S beams, where the visible cracks were primarily vertical.

At higher tension reinforcement, in the NAC-1.32S beam, flexural cracks extended over a longer length. At failure, diagonal flexure-shear cracks progressed at a steep angle towards the compression zone, differing from the NAC-1.32A beam, which displayed a crack pattern similar to NAC-0.75A. The impact of increased reinforcement was evident in the emergence of additional horizontal splitting cracks near the support. A similar pattern of crack growth, influenced by the presence of stirrups, was observed in beams with higher reinforcement areas across each concrete type. Relatively flatter diagonal cracks intersecting more number of stirrups, were observed in the RFA20-1.32S and RFA30-1.32S beams, indicating greater shear resistance contribution from stirrups compared to the NAC-1.32S beams. This relationship between the shear contribution of stirrups and the critical crack shape is consistent with the previous studies on crack kinematics in conventional concrete (Campana et al. 2013). RFA20-1.32S and RFA30-1.32S beams exhibited a critical diagonal crack with more branching, ultimately leading to collapse of the specimens. While stirrups provide more shear resistance than in NAC beams, still they were insufficient to fully control the diagonal crack opening. Closer stirrup spacing is recommended for FARAC beams, though additional studies are needed to establish optimal spacing requirements.

8.3.3 Applicability of Existing Design Provisions

The nominal shear strength of reinforced concrete beams was assessed based on the provisions of multiple standards, including ACI-318, EC 2, CEB-FIP model code 2010 (FIB 2010) and IS-456 (2000). Shear strength, expressions related to compressive strength and the longitudinal reinforcement percentage, as per IS-456 and ACI-318 are shown in Eqs. 8.23 and 8.24 respectively. Equations 8.25 and 8.26 provides shear strength formulations from the CEB-FIP and EC-2 standards, which include an additional parameter a/d or d, to account for the size effect. Notably, the nominal shear capacity of RC beams is calculated without applying partial safety factors. Figure 8.23, 8.24, 8.25 illustrates the results from a database, comprising 100 data points from natural aggregate concrete beams, sourced from Collins et al. (2003). The mean curve fit, obtained using bi-square fitting method, and the 95% upper and lower confidence bounds are shown for the NAC data points. Code based equations are plotted alongside the shear database, which also includes 47 tests data points of RAC beams and fly ash incorporated RAC beams, as reported by Sunayana and Barai (2020).

$$\frac{V_c}{bd\sqrt{f_{cu}}} = \frac{0.85\left(\sqrt{1+5\beta}-1\right)}{6\beta} \tag{8.23}$$

where, $\beta = max\left\{\begin{matrix} \frac{f_{cu}}{6.89\rho} \\ 1 \end{matrix}\right\}$

$$\frac{V_c}{bd\sqrt{f_c}} = \left(1 + 120\rho\frac{Vd}{M\sqrt{f_c}}\right)\frac{1}{7} \le 0.3 \tag{8.24}$$

$$\frac{V_c}{bd{f_c}^{1/3}} = 0.15\left(1+\sqrt{200/d}\right)\left(\frac{3d}{a}\right)^{1/3}\rho^{1/3} \tag{8.25}$$

$$\frac{V_c}{bd{f_c}^{1/3}} = 0.18\left(1+\sqrt{200/d}\right)\rho^{1/3} \tag{8.26}$$

where f_c = cylinder strength; f_{cu} = cube strength; b = width of beam; d = effective depth of beam; ρ = tension reinforcement (%). All the terms in above equations are in SI units.

Figure 8.23 reveals that the Indian standard IS-456 effectively estimates shear capacity across concrete types. Only a few data points corresponding to NAC beams

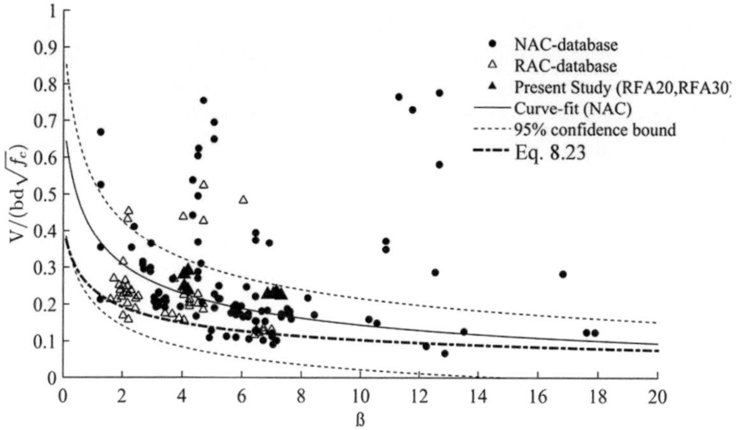

Fig. 8.23 Comparison of experimental database with shear strength prediction of IS-456

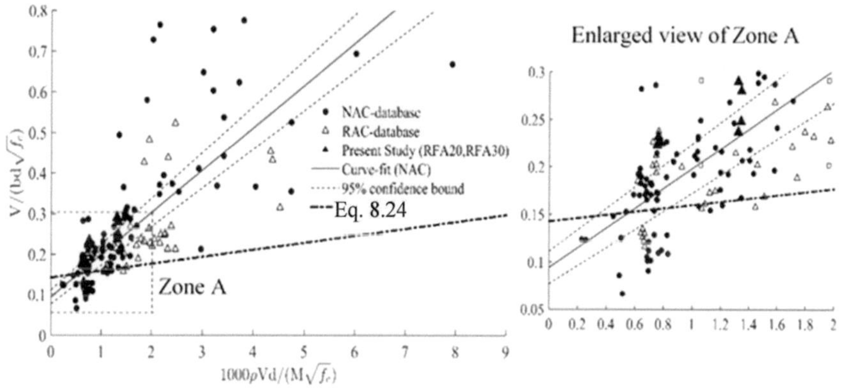

Fig. 8.24 Comparison of experimental database with shear strength prediction of ACI-318

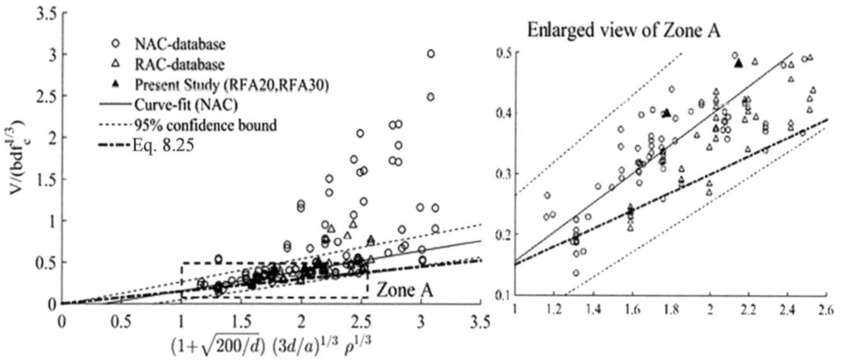

Fig. 8.25 Comparison of experimental database with shear strength prediction of CEB-FIP

failed according to the prediction equation, while RAC beams data align closely with the prediction using Eq. 8.23. This suggests that minor data variability could potentially lead to an unsafe shear strength assessment. Although predictions for fly ash-incorporated RAC beams are somewhat conservative, they remain within the 95% upper confidence limit for NAC. Equation 8.23 tends to underestimate the shear capacity of a few NAC and RAC beams, as indicated by points lying outside the 95% confidence interval. This conservative estimation may stem from the assumption that diagonal cracking load equals failure load. However, the redistribution of stresses after diagonal crack initiation leading to an increase in peak load and ultimate displacement, was observed in case of RFA20 and RFA30 beams (as described in the previous section).

Figure 8.24 compares shear capacity prediction from ACI-318 for NAC, RAC and FARAC beams without shear reinforcement. Equation 8.24 plotted which shows considerable deviation from the fitted curve of the NAC database. Closer examination of 'zone A' reveals that the code prediction for shear may be unsafe for certain NAC and RAC beams with lower $\rho Vd/M\sqrt{f_c}$ value. This observation suggests that the ACI-318 prediction may be unreliable for slender beams with higher a/d ratio at constant main reinforcement ρ and concrete strength f_c. For beams with identical dimensions and concrete properties, increased ρ correlates with safer normalized shear strength. Notably, the RFA20 and RFA30 beams exhibit well-predicted (though highly conservative) normalized shear strength using Eq. 8.24.

Figure 8.25 presents experimental shear database alongside the CEB-FIP model code. The normalized shear strength with respect to $f_c^{1/3}$ tends to be conservatively estimated for the FARAC beams. Equation 8.25 underestimates the shear capacity of RC beams (NAC and RAC data points) with similar cross-sectional dimensions and tension reinforcement (ρ) but smaller shear span, as indicated by some data points beyond 95% upper confidence bound. An examination of the enlarged region 'zone A' reveals that longer shear span or lower ρ in beams with constant cross-sections result in unsafe code predictions for some of the NAC and RAC data points. However, these points lie within 95% lower bound of the mean curve fitted to the NAC database. The EC-2 prediction, plotted in Fig. 8.26, is closer to the mean fit of the NAC beams. For RC beams of higher depth but designed with equivalent ρ, it is observed that experimental shear capacites are lower than predicted, which is even more pronounced in RAC beams.

The applicability of the standard provisions designed for beams made of NAC, as discussed earlier, to RAC and fly ash incorporated RAC beams is further underlined by the ratio of experimental to the predicted values of shear capacity as represented in Table 8.5. Highly over conservative code predictions are observed at lower a/d ratio for RAC beams (Han et al. (2001); Etxeberria et al. (2007)). Further, the replacement of 100% RCA in concrete (Collins et al. 2008; Domingo-Cabo et al. 2009, EN-1992–1–1 2004, Sunayana and Barai 2017) has shown less conservative or even unsafe prediction due to a Additionally, higher tension reinforcement leads to safer prediction of shear capacities of RC beams (Corinaldesi and Giacomo 2009). Table 8.5 provides mean values and the coefficient of variations (COV) according to different code predictions. Euro code yields a closer prediction (mean = 1.14) with

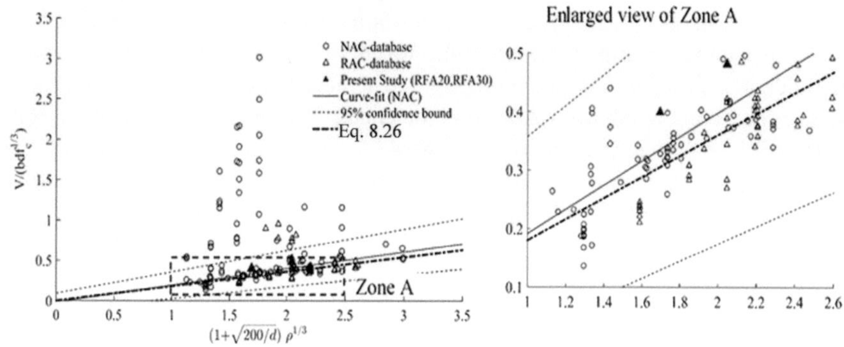

Fig. 8.26 Comparison of experimental database with shear strength prediction of EC-2

a COV of 43.3%, indicating its suitability. On the other hand, the Indian standard (IS-456) is highly conservative (mean = 1.81). The simplified modified compression field theory (SMCFT) was found to be unsafe in estimating the shear capacity of RAC beams (mean ratio = 0.97). Interestingly, for 20% and 30% fly ash incorporated RAC, the estimation of shear capacity according to these provisions satisfied well. The ratio (V_{EXPT}/V_{Code}) followed a similar trend as observed in case of RAC database; IS-456 predictions being overly conservative (mean = 2.12) while other standard methods estimates shear capacity appropriately.

8.3.4 *Reliability of Existing Design Provisions*

The application of coarse RCA introduces additional uncertainty due to its inherent variability. Several factors contribute to this variability in RAC, including the adhered mortar layer on the surface of RCA which leads to the formation of two interfaces (one between parent aggregate and attached old mortar and another between old and fresh mortar), source of the parent concrete, moisture content, and the recycling processes. When the standard shear design provisions of reinforced concrete beams, originally formulated for conventional concrete, are applied to RAC, it may lead to unsafe conditions.

To address this issue, existing studies in this domain have focused on the reliability of bending and shear resistance models specified in major standards. Equivalent partial safety factor was suggested for NAC and RAC beams designed using bending resistance models of American concrete institute (ACI-318) and Eurocode 2. Pacheco et al. (2019) suggested the need for greater partial safety factors for RAC by conducting a preliminary analysis of the shear resistance models in Eurocode and model code. However, it was emphasized that a detailed reliability analysis of limit state of shear is needed. These limited studies underlined the need of calibrating the code-based equations to ensure the safety of the beam elements under shear.

Sunayana and Barai (2021) determined the failure probability of RAC beams, both with and without fly ash, considering the variability associated with it. Necessary modifications to safety factors are proposed to achieve the same reliability as that of NAC or the target reliability specified in the codes. This approach justifies the very matter of the fact that, any new recycled waste materials, in order to be used as one of the constituents in concrete, needs to ensure safety or in other words reliability (with necessary design modifications if required) as that of the original material. The

Table 8.5 Results of experiment to code prediction values of shear force for RAC and FARAC

Author	V_{EXPT} (kN)	$\frac{V_{EXPT}}{V_{IS}}$	$\frac{V_{EXPT}}{V_{ACI}}$	$\frac{V_{EXPT}}{V_{EC}}$	$\frac{V_{EXPT}}{V_{CEB-FIP}}$	$\frac{V_{EXPT}}{V_{SMCFT}}$
Han et al. a/d = 1.5, 2, 3, 4	144.13	3.21	2.94	2.74	2.61	2.32
	117.50	2.62	2.73	2.23	2.34	1.89
	55.08	1.22	1.26	1.05	1.26	0.89
	50.95	1.13	1.15	0.97	1.28	0.82
Etxeberria RCA = 25%, 50% and 100%	104	1.58	1.49	1.05	1.31	0.97
	89	1.35	1.29	0.91	1.13	0.84
	84	1.28	1.24	0.87	1.08	0.80
Choi a/d = 1.5, 2.5, 3.25 RCA = 30%, 50%, 100%	161.7	3.28	2.08	2.09	2.00	1.77
	81.34	1.65	1.26	1.05	1.19	0.89
	56.37	1.14	0.94	0.73	0.90	0.62
	152.88	3.10	1.98	1.99	1.90	1.68
	87.9	1.78	1.38	1.14	1.30	0.97
	71.54	1.45	1.20	0.93	1.15	0.79
	107.8	2.21	1.43	1.44	1.37	1.22
	84.77	1.74	1.36	1.13	1.28	0.96
	57.77	1.18	1.00	0.77	0.95	0.65
Fathifazl a/d = 1.59, 2, 2.59, 3.93 ρ (%) = 1, 1.5, 1.62, 2.46	186.7	4.28	2.74	2.75	2.67	2.27
	169.5	3.30	2.39	2.18	2.28	1.86
	103.9	1.97	1.53	1.27	1.45	1.12
	83.2	1.35	1.28	0.89	1.18	0.81
Arzeoumandi d (mm) = 400, 375, ρ (%) = 1.27, 2.03, 2.71	117.5	2.52	1.07	0.96	1.11	0.75
	151.3	2.72	1.37	1.07	1.29	0.92
	171.8	2.79	1.47	1.10	1.33	0.98
	111.7	2.36	0.97	0.85	1.02	0.68
	148.6	2.63	1.29	1.02	1.22	0.86
	168.7	2.69	1.38	1.05	1.26	0.92
	114.8	2.48	1.07	0.93	1.11	0.75
	143.2	2.59	1.33	1.04	1.25	0.88

(continued)

Table 8.5 (continued)

Author	V_{EXPT} (kN)	$\frac{V_{EXPT}}{V_{IS}}$	$\frac{V_{EXPT}}{V_{ACI}}$	$\frac{V_{EXPT}}{V_{EC}}$	$\frac{V_{EXPT}}{V_{CEB-FIP}}$	$\frac{V_{EXPT}}{V_{SMCFT}}$
	131.4	2.16	1.15	0.86	1.04	0.76
	113	2.40	0.99	0.87	1.05	0.69
	124.1	2.21	1.09	0.86	1.04	0.72
	140.3	2.26	1.17	0.89	1.06	0.77
Knaack and Kurama RCA% = 50, 100	44	0.89	1.46	1.06	1.39	0.96
	39.1	0.79	1.30	0.95	1.24	0.85
	43.7	0.90	1.53	1.09	1.43	0.99
	41.2	0.85	1.44	1.03	1.35	0.94
	36.4	0.75	1.25	0.90	1.18	0.81
	38	0.78	1.30	0.94	1.23	0.85
	39.9	0.82	1.37	0.99	1.29	0.89
	39.9	0.82	1.36	0.99	1.29	0.89
Gonzalez Fontabea	90.64	1.38	1.33	0.93	1.16	0.86
Rahal and Alfrei RCA% = 10, 20, 35, 50, 75, 100	44.5	1.14	0.82	0.81	0.97	0.63
	40.05	1.03	0.75	0.73	0.88	0.58
	45.05	1.16	0.84	0.82	0.99	0.65
	46.95	1.19	0.85	0.84	1.01	0.66
	47.4	1.21	0.87	0.86	1.03	0.67
	42.5	1.08	0.79	0.77	0.93	0.61
Mean values (COV %)		1.81 (47.4)	1.36 (34.3)	1.14 (43.3)	1.31 (31.9)	0.97 (42.8)
Sunayana and Barai Fly ash variation = 20% and 30% ρ (%) = 0.75 and 1.32	82.15	2.14	1.40	1.00	1.15	1.24
	84.45	2.20	1.45	1.04	1.19	1.28
	81.80	2.14	1.42	1.01	1.16	1.25
	80.30	2.10	1.41	1.00	1.15	1.24
	101.73	2.11	1.61	1.04	1.19	1.34
	107.11	2.22	1.67	1.08	1.24	1.39
	87.49	1.81	1.36	0.88	1.01	1.14
	90.15	1.87	1.42	0.92	1.05	1.18
Mean values (COV %)		2.12 (6.7)	1.49 (7.6)	1.01 (5.9)	1.16 (5.9)	1.286.1

following section assess reliability of standard design formulas for the limit state of shear.

8.3.4.1 Methodology

The evaluation of reliability involves following key steps: (i) treatment of database for determination of the model factor (B_{sh}) (ii) identification of random variables involved in the analysis along with their statistical characterization, (iii) definition of limit states, and Monte-Carlo simulations to assess the probability of failure and (iv) modification of resistance factors as necessary to achieve the desired level of reliability, which can be equivalent to NAC or the code-specified target reliability.

The direct effect of coarse RCA on the applicability and accuracy of the design shear resistance models in the considered standards can be quantified known as the model factor. The model factor is defined as the ratio of experimental shear strength to the model predicted value and accounts for the uncertainty associated with the model prediction. B_{sh} can be associated as a multiplicative term in the resistance equation, as suggested by Holicky et al. (2016). It is also worth mentioning that, with variation in material qualities (due to changing raw material production methods and other performance requirements) over time, obtained model factors from old experimental data may not ensure safety. Verification is needed for both economical and safe design.

8.3.4.2 Treatment of Database for Model Factor Estimation

The database comprised of shear-critical NAC and RAC beam specimens designed without stirrups in the shear span and which failed under shear. The NAC, RAC and FARAC database was created from the experimental results of Torsak et al. (2001), Yang et al. (2003) Lubell et al. (2004), Rahal and Al-Shaleh (2004), Sherwood et al. (2006), Tariq and Newhook (2003) Yun et al. (2011), Katkhuda and Shatarat (2016), Sunayana and Barai (2020) and Sunayana and Barai (2021). The database is briefly summarized in Table 8.6. Only beam specimens with a rectangular cross-section and normal weight aggregates in the concrete mix were considered. There were no specific limits on the compressive strength of concrete or dimensions of specimens, although deep beams were excluded. Initially, 143 number of NAC and 115 RAC (with and without fly ash) samples were considered. The theoretical shear capacity was calculated using the shear resistance models from IS-456, ACI-318, CEB-FIB model code and EC-2 as discussed in the previous section. Material safety factor terms used in these equations were assumed as unity to predict actual capacity. The ratio of the experimental shear strength to the corresponding theoretical shear was defined as model factors (B_{sh}).

The method suggested by Holicky et al. (2016) was followed for treatment of the database to find out model factors. It involved four steps as follows: (i) identification of outliers (ii) verification of randomness (iii) fitting of suitable distribution curve to the existing database and (iv) calculation of statistical properties of model factors. Outliers were identified and removed from the database based on specific criteria as illustrated in Fig. 8.27. Dispersion of the data were plotted where, the vertical lines along with the interquartile range (IQR) i.e. 25–75 percentile depicted the

Table 8.6 Summary of the database used for calculation of shear model factor

Parameters	NAC	RAC and FARAC
Width (b) in mm	$101 \leq b \leq 3000$	$150 \leq b \leq 400$
Effective depth (d) in mm	$166 \leq d \leq 1067$	$150 \leq d \leq 525$
Shear span to depth ratio (a/d)	$2.3 \leq a/d \leq 5.1$	$2.5 \leq a/d \leq 5.1$
Tension reinforcement (ρ) in %	$0.44 \leq \rho \leq 3.89$	$0.53 \leq \rho \leq 2.98$
Compressive strength of concrete (f_c) in MPa	$18.6 \leq f_c \leq 68.9$	$18.1 \leq f_c \leq 47.8$
Replacement (r) in %	100	$10 \leq r \leq 100$

maximum and minimum values. Values of model factors beyond a specified range (1.5 times IQR and greater than 3 times the standard deviation of a particular dataset) were considered potential outliers and were removed accordingly. The same process was applied to both NAC and RAC databases, resulting in mean and coefficient of variation calculations for model factors for different standards. Total 143 number of NAC samples were initially considered out of which 130 samples were sorted following the outlier removal method explained above. For RAC, total 107 numbers of data points were considered.

The second step involved the verification of the model factor as an independent random variable, and this process is illustrated for a typical case of ACI shear resistance model in Fig. 8.28. Similar assessment of model factors related to other resistance models were conducted. The correlation between the model factor and other variables such as b, d, f_c, ρ and a/d was evaluated using the bivariate Pearson and Spearman correlation coefficients. Values of these coefficients closer to 0 indicated insignificant correlation between the model factors and different variables. This exercise justified considering the model factor as an independent random variable. The third step aimed to find out the probabilistic distribution of model factor variable. Figure 8.29 represents the histogram count along with a normal distribution plot for a typical case of ACI model factors, calculated using both NAC and RAC shear database. The distribution function fitted to the model factors based on other

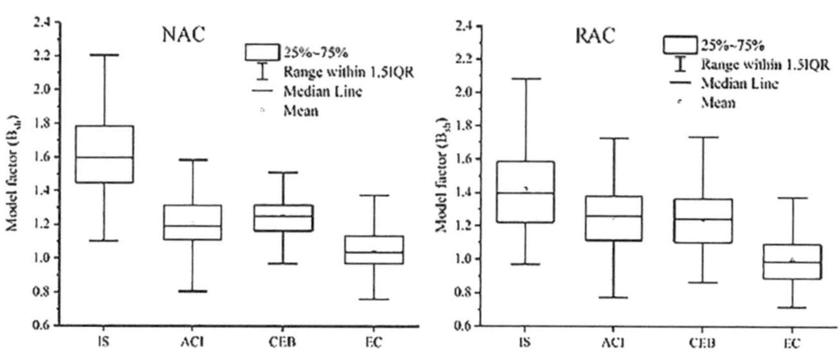

Fig. 8.27 Statistical treatment of database for calculation of shear model factors

standards was also examined. To ensure normality of the model factor, the Anderson darling test was performed, following the method outlined by Dithinde et al. (2011). The p value calculated from the test was found to be 0.604, which was greater than the significance level of 0.05. This result justified the acceptance of the hypothesis that the model factors follow a normal distribution. A similar process was repeated for model factors based on other standards.

The final step in the treatment of the database involved the statistical characterization of the variable B_{sh}. It is important to note that model factors obtained directly from the comparison of experimental to theoretical models also contain variability stemming from experimental testing methods (including measurement accuracy of strain and displacement) specimens size effect (parameter relating the laboratory results to real structural behaviour) and time-variant effects. So, coefficient of variation (COV) of model factors were calculated by removing the variability attributed to testing and specimen size effect (considered as 4%) using the following Eq. 8.27 (Mirza et al. 1979).

$$COV_{model} = \sqrt{COV_{test/model}^{2} - COV_{test}^{2} - COV_{specimen}^{2}} \tag{8.27}$$

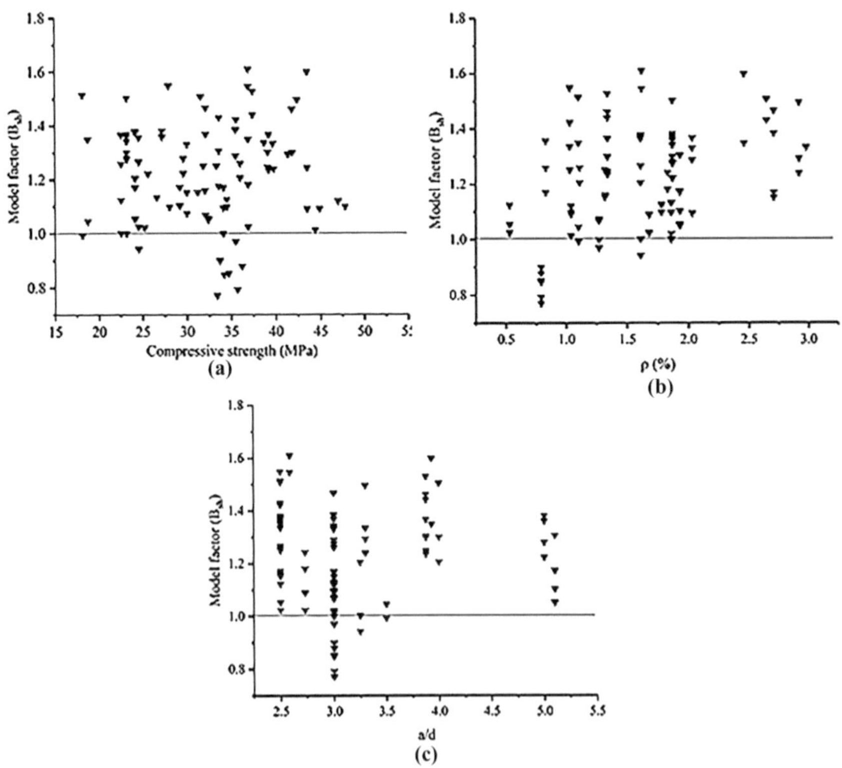

Fig. 8.28 Scatter plots of ACI shear model factor versus variables **a** f_c **b** ρ and **c** a/d

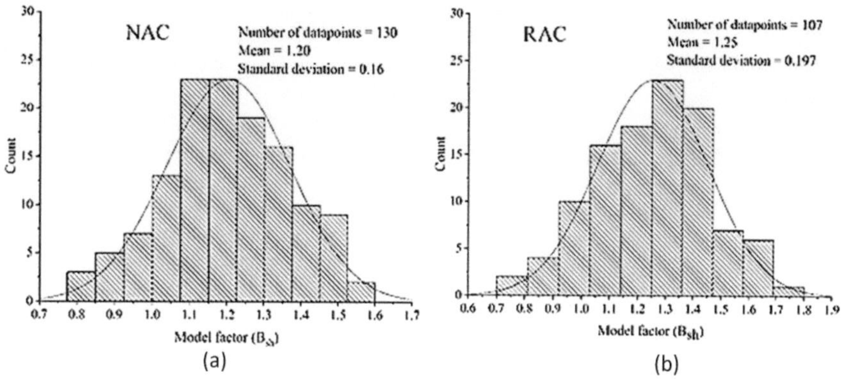

Fig. 8.29 Distribution of model factors **a** NAC database **b** RAC database

In reliability analysis, it is important to identify the random variables which are the sources of variability and quantify them in terms of statistical properties such as mean, COV and their probabilistic distribution. The next section focusses on basic random variables for designing under limit state of shear.

8.3.4.3 Basic Random Variables

Table 8.7 provides an overview of basic variables and their statistical properties including bias and the COV. The bias of a random variable was calculated as the ratio of its mean value to the nominal value. This approach of using the bias value helps minimizing the effect of nominal value, making the results applicable under various design conditions.

Load variables: The stochastic live load parameter was assumed to follow extreme type 1 distribution. Permanent dead load was presumed to follow normal distribution. To assess different design scenarios, the live load (LL) to dead load (DL) ratio (LL/DL) was varied. All the design cases used the same load factors for dead and live loads, but the ratio between magnitudes of dead load and live load intensity varied. The live load to dead load ratio (LL/DL) considered in this study were 0.5, 1, 1.5, 2, 3, and 4.

Concrete strength: The actual strength of concrete may deviate from the characteristic strength due to various factors like material quality, weighting, mixing method, placing, compaction, curing conditions, transportation method and testing procedure etc. (Ranganathan 1990). In this study, statistical properties of the compressive strength of concrete were considered based on data from Xiao et al. (2017) for both the RAC and NAC. It was assumed that the compressive strength follows a normal distribution (Ranganathan 1990). The mean and COV of compressive strength in RAC were slightly lower compared to NAC due to the greater variation associated with sources of parent concrete for the former case.

Table 8.7 Properties of random variables for limit state of shear

Random variables	Bias	COV (%)
Dead load	1.05	11
Live load	1.0	25
Width of beam	1.01	4
Effective depth of beam	1.0	4
Area of steel	1.0	4
NAC		
Compressive strength Kou and Poon (2012)	1.39	8.3
Model factor		
$B_{sh,ACI}$	1(mean = 1.22)	13.69
$B_{sh,IS}$	1(mean = 1.40)	14.02
$B_{sh,CEB}$	1(mean = 1.25)	10.79
$B_{sh,EC}$	1(mean = 1.05)	10.00
RAC		
Compressive strength Kou and Poon (2012)	1.31	9.2
Model factor		
$B_{sh,ACI}$	1(mean = 1.25)	14.64
$B_{sh,IS}$	1(mean = 1.24)	16.64
$B_{sh,CEB}$	1(mean = 1.22)	12.66
$B_{sh,EC}$	1(mean = 0.99)	13.6

Dimensions of beam: Geometrical parameters of beam cross-sections, such as width and effective depth, were considered as normally distributed random variables (Lu et al. 1994). Randomness of these geometrical quantities primarily arises from the variations in quality, control.

From the Table 8.7 it can be noted that, bias values of dimensional and load variables are smaller; these values closer to 1 implied that the mean tends towards its nominal value. However, bias of concrete strength as well as model factors are relatively higher than 1 which indicated higher uncertainty and significant difference between the mean and its nominal value. The random variables with higher bias and COV are believed to affect reliability of the shear resistance models significantly.

8.3.4.4 Reliability Assessment

A simply supported beam with cross-sections 200 mm width and 300 mm overall depth was designed to withstand the load combination prescribed in the standards. Table 8.8 shows the code-specified load cases and resistance factors considered for design of RC beams. To assess the shear resistance, Monte-Carlo simulations were performed to generate all the random variables involved in the calculation of shear resistance. Then the shear limit state function (G) was formulated using the resistance function R and load function L as G (X) = R (X)—L (X); X denotes a vector of influencing random variables. The condition satisfying G < 0 indicated failure and consequently probability of failure was calculated. Initially, the code prescribed resistance factors were considered to determine the reliability index (ß) of NAC and RAC beams under limit state of shear.

Finally, the resistance factors were modified based on the database RAC beams with and without fly ash in order to achieve the target reliability level for limit state of shear. The CEB standard specified a reliability index of 3.8 for NAC for 50 years' service life of the structure and a medium consequence of failure. The target reliability for RAC beams was set as the smaller value between this specified standard value and the reliability of NAC calculated using the procedure discussed earlier. This allowable probability of failure, as set by CEB standard, takes into account factors such as the desired service life of the structure, cost of safety and other consequences of failure such as injury, life-risk and economic losses etc.

The deterministic nominal resistance values were calculated initially based on the shear resistance equations. Then the statistical properties of basic random variables affecting shear were generated through Monte-Carlo simulations. Each simulation involved fifty million sample points and was executed using MATLAB environment. The probability of failure was estimated for various design cases, characterized by a different LL/DL ratio, such as 0.5, 1, 1.5, 2, 3 and 4. The reliability index (ß) corresponding to each probability of failure value was determined separately for NAC and RAC. The resistance factor for RAC was modified accordingly to achieve the target reliability. It is noteworthy that, CEB, EC and IS standards consider material partial safety factors in their design. However, the reliability analysis resulted very high material safety factor due to the form of equation (f_c is raised to the power 0.5 or 0.33). Hence, the load resistance factor design method was used in all the cases to determine resistance factors. Figure 8.30 illustrates variation of ß with LL/DL ratio for RC beams designed in shear. ß decreases with increasing LL/DL ratio.

Based on the ACI provision, ß values were obtained as 3.7 to 3.96 in NAC whereas 3.58 to 3.74 with the average 3.66 for RAC. Similarly, IS provisions resulted in reliability index in the range of 4.4 to 3.86 for NAC and 3.47 to 3.16 for RAC. Reinforced concrete beams designed with IS code method showed significant difference between the reliability of RAC and NAC beams with much larger probability of failure in the earlier case. It may be due to over-conservative estimation of aggregate interlocking in resisting shear; this component is lower in the case of RAC beams because of

Table 8.8 Code-specified load combinations and resistance factors

Standard provisions	Load combination	Resistance or material factor
IS	1.5DL + 1.5LL	0.8
ACI	1.2DL + 1.6LL	0.75
CEB	1.35DL + 1.5LL	0.8
EC	1.35DL + 1.5LL	0.8

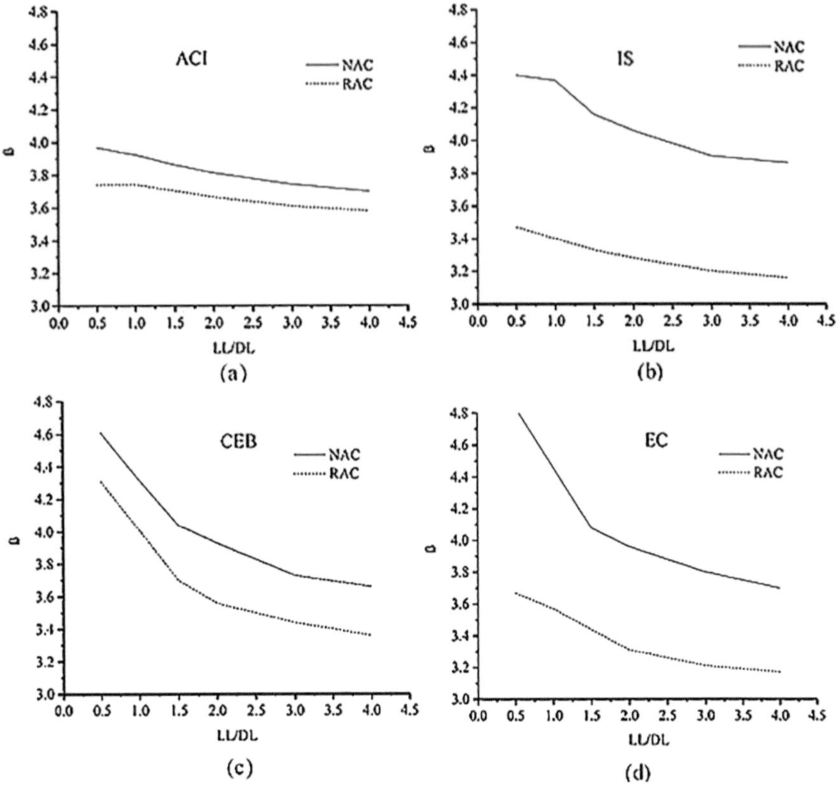

Fig. 8.30 Variation of reliability index with live load to dead load ratio **a** ACI **b** IS **c** CEB **d** EC

smooth surface texture of RCA providing lesser frictional resistance. ß calculated with CEB-FIP were ranged from 4.61 to 3.66 for NAC and 4.31 to 3.36 for RAC. EC-2 standard resulted in reliability index of 4.84 to 3.7 for NAC and 3.67 to 3.17 for RAC. Lower reliability of RAC under shear limit state may be attributed to higher variability of model factors as well as compressive strength of concrete. Code calibration was carried out to achieve the same reliability of RAC and NAC beams with the same design provisions in shear. Resistance factors are suggested for RAC beams to attain acceptable level of safety under limit state of shear. Table 8.9 shows the modified resistance factors for different cases of LL/DL ratio varying from 0.5 up to 4. It can be observed that, the revised resistance factor for ACI, IS, CEB and EC standards varied respectively in the range of 0.7–0.72, 0.69–0.62, 0.8–0.72 and 0.66–0.56 respectively.

Table 8.9 Shear resistance factor for RAC

LL/DL	Shear resistance factors of RAC			
	ACI	IS	CEB	EC
0.5	0.7	0.697	0.8	0.66
1	0.7193	0.68	0.8	0.6326
1.5	0.7105	0.66	0.78	0.6109
2	0.7243	0.6503	0.7697	0.6025
3	0.7263	0.62	0.7419	0.5677
4	0.7184	0.615	0.72	0.5604

8.4 Summary

In this chapter, flexural behaviour of FARAC beams was investigated emphasizing the effect of tension reinforcement percentage, compression reinforcement and concrete type. Load–deflection curves and failure modes were observed. Applicability of code provisions and normalized moment capacity were assessed through comparative study with the available database of RAC. Tension-stiffening of FARAC was evaluated numerically. The results obtained from the study can be summarized below:

- At ultimate limit state, the flexural performance of RAC incorporating fly ash reinforced concrete beam designed with PPM mix design is equivalent as compared to the conventional concrete beam. The existing code provisions for NAC beams can be applied to RAC incorporating fly ash beam irrespective of tension reinforcement percentage.
- The maximum mid-span deflection for RAC with fly ash reinforced concrete beam is higher than NAC beam, but the variation is not very significant at service load. In addition, the deflection is well within the maximum allowable limit of ACI 318, IS 456 (more conservative) and Eurocode 2 provisions.
- The higher mid-span deflection can be attributed to lesser tension stiffening effect in RAC mostly as a consequent of lower tensile strength. The constitutive relation for cracked tensile concrete is obtained for both NAC and RAC. Higher mid-span deflection can be attributed to lesser tension stiffening effect in RAC as a consequent of lower tensile strength. Constitutive relation was obtained for cracked tensile concrete from the experimental moment–curvature curves for both NAC and FARAC. Serviceability reliability of RAC beams were analysed for limit state of deflection. Necessary modification of span-to depth ratio was suggested for RAC in order to obtain NAC-equivalent or code-specified target reliability.
- Many numbers of closely spaced cracks are observed for RAC with fly ash beam which is even more significant at higher reinforcement percentage. By using particle packing method of mix design, it is possible to produce concrete with 100% recycled aggregates along with up to 30% fly ash substitution without compromising flexure performance of reinforced concrete beam.

The shear performance of RC beams comprising fly ash incorporated recycled aggregate (100%) concrete are evaluated following the particle packing mix design method. The effect of aggregate types and cementitious materials (fly ash in 100% RCA), tension reinforcement percentage and presence of stirrups are examined. Behaviour of FARAC under shear can be summarized below.

- Fly ash incorporation (up to 30%) in 100% recycled coarse aggregate based concrete beams showed equivalent (within 10% variation) shear strength at low tension reinforcement (ρ) as compared to natural aggregate concrete beams. However, in presence of higher ρ, failure shear reduced by about 20%. The concrete properties such as tensile strength and failure pattern in FARAC affect the shear carrying components adversely; the effect is more significant at higher tension reinforcement.
- Analysis of location and angle of critical crack highlights the contribution of shear carrying mechanisms in case of FARAC beams without stirrups. The less inclined critical cracks demonstrate poor aggregate interlocking. Lesser contribution of dowel action is indicated by formation of critical crack at larger distance from the support. Critical cracks intercepting more number of stirrups signify greater stirrups contribution in RFA20 and RFA30 beams as compared to NAC.
- Statistical analysis of ACI-318, EC-2, CEB-FIP model code, IS-456 and SMCFT methods underlines over conservativeness of IS provisions and closer prediction of EC2 for the RAC data points from pr and the results of FARAC beams obtained from the present work.
- Safe applicability of the existing shear provisions to beams constituted of 100% RAC (with or without fly ash) require higher tension reinforcement and lower a/d ratio. Otherwise, proper consideration of RCA replacement effect is recommended for computation of shear capacity of beams without stirrups.

The significant reduction of failure shear and higher critical crack width in 30% fly ash incorporated RAC can be mitigated by providing sufficient shear reinforcement. Further investigations regarding exact requirement of shear reinforcement and higher fly ash replacement (beyond 30%) in 100% RAC are required for widespread structural applications of this sustainable construction material.

Probability of shear and serviceability failure of reinforced concrete beam comprising recycled coarse aggregates were estimated. Effect of randomness associated with concrete material properties, dimensions, load variables and model uncertainty were assessed. Monte-Carlo simulation was performed for evaluation of statistical properties of random variables and reliability analysis. Based on this study, following conclusions can be derived.

- Model factors indicated slight un-conservative application of Eurocode to design RAC beams. Mean values of model uncertainty factors calculated using ACI, IS, CEB model code and EC were mostly similar; but variability was higher for RAC as compared to NAC. Due to this effect as well as higher coefficient of variation of compressive strength, reliability index was reduced for RAC under limit states of shear.

- Modified shear resistance factors were proposed to achieve NAC equivalent and code-specified safety of the structures. The shear resistance (or material) factors existing for NAC are 0.8 as per IS, CEB and EC and 0.75 as per ACI standards. For a LL/DL ratio of 0.5 to 4, the revised resistance factors for RAC were obtained in the range of 0.70–0.72 for ACI, 0.69–0.62 for IS, 0.8–0.72 for CEB and 0.66–0.57 for EC.
- Probability of failure for deflection was higher for RAC on account of higher variability associated with the influencing parameters. The span-to effective depth ratio was revised for RAC as 12.5 to 16.6 (up to 10 m span) and 17.5 and 18.2 for span 12 m and 14 m respectively. The Australian code equation was found suitable for calculation of l/d ratio for RAC without any alterations.

The recommended shear resistance factors and span-to effective depth provisions will ensure adequate allowable safety levels as per major standards for RAC beams. This study will be beneficial towards gaining confidence for wider application of RAC. Moreover, further investigations are required for different boundary conditions, loading and other variability.

References

ACI 318–11 (2011) Building code requirements for structural concrete and commentary. American Concrete Institute, Farmington Hills, Michigan, USA

Agrela F, Alaejos P, de Juan MS (2013) Properties of concrete with recycled aggregates, In: pacheco-Torgal et al (eds) Handbook of recycled concrete and demolition waste, Woodhead, London, UK

Ajdukiewicz AB, Kliszczewicz AT (2007) Comparative tests of beams and columns made of recycled aggregate concrete and natural aggregate concrete. J Adv Concr Technol 5(2):259–273

Arezoumandi M, Smith A, Volz JS, Khayat KH (2014) An experimental study on shear strength of reinforced concrete beams with 100% recycled concrete aggregate. Constr Build Mater 53:612–620

Arezoumandi M, Smith A, Volz JS, Khayat KH (2015) An experimental study on flexural strength of reinforced concrete beams with 100% recycled concrete aggregate. Eng Struct 88:154–162

Al-Zahraa F, El-Mihilmy MT, Bahaa T (2011) Experimental investigation of shear strength of concrete beams with recycled concrete aggregates. Int J Mater Struct Integrity 5(4):291–310

Bai WH, Sun BX (2010) Experimental study on flexural behavior of recycled coarse aggregate concrete beam. Appl Mech Mater 29:543–548

Bažant ZP, Kim JK (1992) Improved prediction model for time-dependent deformation of concrete, Part 3: Creep at drying. Mater Struct 25(1): 21–28. https://doi.org/10.1007/BF02472209

Bentz EC, Buckley S (2005) Repeating a classic set of experiments on size effect in shear of members without stirrups. ACI Struct J 102:832–838

Breccolotti M, Materazzi AL (2013) Structural reliability of bonding between steel rebars and recycled aggregate concrete. Constr Build Mater 47:927–934

Campana S, Ruiz MF, Anastasi A, Muttoni A (2013) Analysis of shear-transfer actions on one-way RC members based on measured cracking pattern and failure kinematics. Mag Concr Res 65:386–404

Choi ECC (1989) Live load model for office buildings. Struct Eng Div 67(24):421–437

Choi WC, Yun HD (2017) Shear strength of reinforced recycled aggregate concrete beams without shear reinforcements. J Civ Eng Manag 23:76–84

Choi WC, Yun HD, Kim SW (2012) Flexural performance of reinforced recycled aggregate concrete beams. Mag Concr Res 64(9):837–848
Collins MP, Bentz EC, Sherwood EG (2008) Where is shear reinforcement required? Review of research results and design procedures. ACI Struct J 105:590–600
Corinaldesi V, Giacomo M (2009) Influence of mineral additions on the performance of 100% recycled aggregate concrete. Constr Build Mater 23:2869–2876
De Brito J, Saikia N (2011) Recycled aggregate in concrete: use of industrial, construction and demolition waste. Springer science and business media
Domingo-Cabo A, Lázaro C, López-Gayarre F, Serrano-López MA, Serna P, Castaño-Tabares JO (2009) Creep and shrinkage of recycled aggregate concrete. Constr Build Mater 23:2545–2553
Ellingwood B, Culver C (1977) Analysis of live loads in office buildings. J Struct Eng ASCE 103(8):1551–1560
Ellingwood B (1978) Reliability basis of load and resistance factors for reinforced concrete design, NBS building science series. Natl Bur Stan. Washington, DC 110(44)
EN-1992–1-1, Eurocode 2—Design of Concrete Structures (2008) Part 1–1: general rules and rules for buildings, London; British Standard Institution 2004
Etxeberria M, Mari AR, Vázquez E (2007) Recycled aggregate concrete as structural material. Material and Structures 40:529–541
Fathifazl G, Razaqpur AG, Isgor OB, Abbas A, Fournier B, Foo S (2009a) Shear strength of reinforced recycled concrete beams without stirrups. Mag Concr Res 61:477–490
Fathifazl G, Razaqpur AG, Isgor OB, Abbas A, Fournier B, Foo S (2009b) Flexural performance of steel-reinforced recycled concrete beams. ACI Struct J 106:858–867
Fathifazl G, Razaqpur AG, Isgor OB, Abbas A, Fournier B, Foo S (2011a) Shear capacity evaluation of steel reinforced recycled concrete (RRC) beams. Eng Struct 33:1025–1033
Fathifazl G, Razaqpur AG, Isgor OB, Abbas A, Fournier B, Foo S (2011b) Creep and drying shrinkage characteristics of concrete produced with coarse recycled concrete aggregate. Cement Concr Compos 33(10):1026–1037
FIB Model code for concrete structures (2010) Ernst and Sohn
Gastebled OJ, May IM (2001) Fracture mechanics model applied to shear failure of reinforced concrete beams without stirrups. Struct J 98(2):184–190
Gilbert RI (1988) Time effects in concrete structures. Elsevier, Amsterdam, The Netherlands
González-Fonteboa B, Martínez-Abella F (2007) Shear strength of recycled concrete beams. Constr Build Mater 21:887–893
Gribniak V (2009) Shrinkage effect on tension-stiffening of concrete structure. Dissertation, Vilnius Gediminas Technical University, Vilnius
Han BC, Yun HD, Chung SY (2001) Shear capacity of reinforced concrete beams made with recycled-aggregate. ACI Spec Publ 200:503–516
Huber P, Huber T, Kollegger J (2016) Investigation of the shear behavior of RC beams on the basis of measured crack kinematics. Eng Struct 113:41–58
Harris ME, Corotis RB, Bova CJ (1981) Area-dependent processes for structural loads. J Struct Eng 107:857–872
Ignjatović IS, Marinković SB, Mišković ZM, Savić AR (2013) Flexural behavior of reinforced recycled aggregate concrete beams under short-term loading. Mater Struct 46(6):1045–1059
Ignjatović IS, Marinković SB, Tošić N (2017) Shear behaviour of recycled aggregate concrete beams with and without shear reinforcement. Eng Struct 141:386–401
IS: 456 (2000) Plain and reinforced concrete code of practice, Bureau of Indian Standards, New Delhi, India
Kaklauskas G (2004) Flexural layered deformational model of reinforced concrete members. Mag Concr Res 56:575–584
Kaklauskas G, Gribniak V, Bacinskas D, Vainiunas P (2009) Shrinkage influence on tension stiffening in concrete members. Eng Struct 31(6):1305–1312
Kang THK, Kim W, Kwak YK, Hong SG (2014) Flexural testing of reinforced concrete beams with recycled concrete aggregates. ACI Struct J 111(3)

Karshenas S, Ayoub H (1994) Analysis of concrete cnstruction live loads on newly poured slabs. J Struct Eng 120:1525–1542

Katkhuda H, Shatarat N (2016) Shear behavior of reinforced concrete beams using treated recycled concrete aggregate. Constr Build Mater 125:63–71

Knaack AM, Kurama YC (2014) Behavior of reinforced concrete beams with recycled concrete coarse aggregates. J Struct Eng (ASCE) 141:B4014009–1–12

Kou SC, Poon CS (2012) Enhancing the durability properties of concrete prepared with coarse recycled aggregate. Constr Build Mater 35:69–76

Kou SC, Poon CS, Chan D (2007) Influence of fly ash as cement replacement on the properties of recycled aggregate concrete. J Mater Civ Eng 19(9):709–717

Lu R, Luo Y, Conte JP (1994) Reliability evaluation of reinforced concrete beams. Struct Saf 14(4):277–298. https://doi.org/10.1016/0167-4730(94)90016-7

Lubell A, Sherwood T, Bentz EC, Collins MP (2004) Safe shear design of large, wide beams. Concr Int 26:66–78

Lye CQ, Dhir RK, Ghataora GS, Li H (2016) Creep strain of recycled aggregate concrete. Constr Build Mater 102:244–259

Mahgoub M, Jamali A, Saadeghvaziri MA (2017) Behavioral model for recycled aggregate concrete under axial compression. Special Publication 314:1–12

Maruyama I, Sogo M, Sogabe T, Sato R, Kawai K (2004) Shear behaviour of reinforced recycled concrete beams. In: Proceedings of the conference on the use of recycled materials in building and structures, Barcelona, Spain

Mirza SA, Hatzinikolas M, MacGregor JG (1979) Statistical descriptions of strength of concrete. J Struct Eng 105(6):1021–1037

Pacheco J, de Brito J, Chastre C, Evangelista L (2019) Uncertainty models of reinforced concrete beams in bending: Code comparison and recycled aggregate incorporation. J Struct Eng 145(4):04019013

Pham L (1985) Load combinations and probabilistic load models for limit state codes. Civ Eng Trans 27:62 67

Pollard JH (1979) A handbook of numerical and statistical techniques. CUP Archive, Cambridge, p 368

Popovics S (1973) A numerical approach to the complete stress-strain curve of concrete. Cem Concr Res 3:583–599

Ranganathan R (1990) Structural reliability analysis and design. Jaico Publishing House

Rahal KN, Alrefaei YT (2017) Shear strength of longitudinally reinforced recycled aggregate concrete beams. Eng Struct 145:273–282

Rahal KN, Al-Shaleh KS (2004) Minimum transverse reinforcement in 65 MPa concrete beams. ACI Struct J 101:872–878

RILEM draft recommendation: TC-242-MDC multi-decade creep and shrinkage of concrete: material model and structural analysis, Model B4 for creep, drying shrinkage and autogenous shrinkage of normal and high-strength concretes with multi-decade applicability (2015) Mater Struct 48:753–770

Sadati S, Arezoumandi M, Khayat KH, Volz JS (2016) Shear performance of reinforced concrete beams incorporating recycled concrete aggregate and high-volume fly ash. J Clean Prod 115:284–293

Sagher AM (2016) Shear strength of reinforced concrete beams made with recycled coarse aggregate, Doctoral dissertation

Seara-Paz S, González-Fonteboa B, Martínez-Abella F, Eiras-López J (2018) Flexural performance of reinforced concrete beams made with recycled concrete coarse aggregate. Eng Struct 156:32–45

Sherwood EG, Lubell AS, Bentz EC, Collins MP (2006) One-Way Shear Strength of Thick Slabs and Wide Beams. ACI Struct J 103:794–802

Silva RV, De Brito J, Dhir RK (2015) Prediction of the shrinkage behavior of recycled aggregate concrete: a review. Constr Build Mater 77:327–339

Stewart MG (1996) Serviceability reliability analysis of reinforced concrete structures. J Struct Eng 122:794–803

Sturm AB, Visintin P, Oehlers DJ (2017) Time-dependent serviceability behavior of reinforced concrete beams: partial interaction tension stiffening mechanics. Structural Concrete 19(2):508–523

Sunayana S, Barai SV (2017) Recycled aggregate concrete incorporating fly ash: Comparative study on particle packing and conventional method. Constr Build Mater 156:376–386

Sunayana S, Barai SV (2018) Flexural performance and tension-stiffening evaluation of reinforced concrete beam incorporating recycled aggregate and fly ash. Constr Build Mater 174:210–223

Sunayana S, Barai SV (2020) Shear behavior of fly-ash-incorporated recycled aggregate concrete beams. ACI Struct J 117(1):289–303

Sunayana S, Barai SV (2021) Shear serviceability reliability of recycled aggregate concrete beams. ACI Struct J 118(2):225–236

Tariq M, Newhook JP (2003) Shear testing of FRP reinforced concrete without transverse reinforcement. In: Annual conference of the Canadian society for civil engineering, Moncton, Canada

Tavakoli M, Soroushian P (1996) Drying shrinkage behavior of recycled aggregate concrete. Concr Int 18(11):58–61

Torsak L, Akinori Y, Maki M, Junichiro N (2001) Concept of concrete compressive fracture energy in RC deep beams without transverse reinforcement. Trans Jpn Concr Inst 23:119–124

Tung ND, Tue NV (2016) A new approach to shear design of slender reinforced concrete members without transverse reinforcement. Eng Struct 107:180–194

Wardeh G, Ghorbel E (2019) Shear strength of reinforced concrete beams with recycled aggregates. Adv Struct Eng 22:1938–1951

Xiao J, Li J, Zhang C (2005) Mechanical properties of recycled aggregate concrete under uniaxial loading. Cem Concr Res 35(6):1187–1194

Xiao J, Falkner H (2007) Bond behaviour between recycled aggregate concrete and steel rebars. Constr Build Mater 21:395–401

Xiao J, Zhang K, Xie Q (2016) Reliability analysis for flexural capacity of recycled aggregate concrete beams. Struct Eng Int 26(2):121–129

Yang KH, Chung HS, Lee ET, Eun HC (2003) Shear characteristics of high strength concrete deep beams without shear reinforcements. Eng Struct 25:1343–1352

Yang Y (2014) Shear behaviour of reinforced concrete members without shear reinforcement: a new look at an old problem. Doctoral dissertation, TU Delft, Delft University of Technology

Yanweerasak T, Kea T, Ishibashi H, Akiyama M (2018) Effect of recycled aggregate quality on the bond behavior and shear strength of RC members. Appl Sci 8(11):2054

Yun HD, You YC, Lee DH (2011) Effects of replacement ratio of recycled coarse aggregate on the shear performance of reinforced concrete beams without shear reinforcement. LHI Journal of Land Housing and Urban Affairs 2(4):471–477. https://doi.org/10.5804/LHIJ.2011.2.4.471

Zhu WQ, Meng G, Jia JQ (2014) Experimental studies on axial load performance of high-strength concrete short columns. Proc Inst Civ Eng-Struct Build 167:509–519

Chapter 9
Behaviour of Reinforced Concrete Slabs Under Punching Shear

Preview

Flat slab supported by an interior column is a complicated problem due to the most likely phenomenon of punching shear failure. This chapter discusses the punching shear behaviour of recycled aggregate concrete with coarse RCA. In the context of using SCMs in concrete, we address the following question: How does low-volume fly ash as a cement substitute alter the punching shear of RAC flat slabs? Another aspect while using any recycled material is maintaining the level of structural safety as that of the already existing material. The punching shear resistance design equation, available for the standard reinforced concrete slabs, may not yield a safe design for RAC slabs due to greater uncertainties. The reliability of standard punching shear models is assessed and modified to ensure equivalent safety for RAC and NAC flat slabs. It is crucial to conduct this reliability analysis in order to confidently establish fly ash and 100% coarse RCA as alternative concrete constituents in practice.

9.1 Effects of Coarse RCA on Punching Shear Behaviour of Slabs

The punching shear strength mainly depends on the compressive strength (or tensile strength) of concrete, flexure reinforcement percentage, shape of the column and its dimensions as well as depth of the slab (Sacramento et al. 2012). In the recent past, few research studies examined the effect of recycled coarse aggregates (RCA) on the punching shear behaviour of reinforced concrete flat slabs. Similar punching strength and failure patterns of RAC (20%, 50%, and 100% RCA) slabs and NAC slabs was reported due to the equivalent compressive strength of concrete (Reis et al. 2015). Stiffness of the specimens and cracking load decreased due to reduction in modulus of elasticity and tensile strength of RAC as compared to NAC. In a similar kind of

S. Sunayana and S. V. Barai, *Green Concrete*, Springer Transactions in Civil and Environmental Engineering, https://doi.org/10.1007/978-981-96-6982-0_9

study (Francesconi et al. 2016), the predominant role of reinforcement rather than concrete on the punching strength of slabs specimens was observed. Shear strength as well as cracking load in RAC (RCA at replacement of 30%, 50%, 80%, and 100%) was also found similar. But, the deflections at cracking and ultimate load were higher. On the contrary, reduction in the cracking as well as the ultimate load and increase in the corresponding deflection at higher replacement of RCA was reported in flat slabs (Mahmoud et al. 2018). Higher RCA also resulted in the lowering of stiffness and energy absorption capacity of concrete slabs. Reduction of punching load with the increasing replacement percentage of RCA can be compensated (improved punching shear capacity as well as ductility) by incorporating steel fibres in concrete (Xiao et al. 2018). In a recent study by Sahoo and Singh (2020), negligible effect of RCA was reported on the punching shear behaviour of flat slabs. The observations from the above-mentioned limited studies do not give a clear picture and sometimes are contradictory as far as the effect of RCA on punching shear is concerned. The design punching shear equations of different standards are mostly empirical in nature because the coefficients of influential parameters are determined using statistical analysis of old database. So a reliability analysis study is discussed in this chapter.

9.2 Effects of Fly Ash and RAC

This section discussed the structural application of 100% RAC and low-volume fly ash (denoted as FARAC) as a cement substitute in reinforced concrete flat slabs under punching shear. Load–deflection, ultimate load and failure pattern of fly ash incorporated RAC slabs are measured and compared with the control NAC slabs.

9.2.1 Slab Specimen Details and Loading Set-Up

A total of six slab specimens; two similar specimens for each concrete type, are designed, fabricated, and tested underlining the effect of RFA20 and RFA30. PPM mix design is adopted for the concrete. Simply supported square flat slab specimens of dimension 1000 mm × 1000 mm × 100 mm are designed according to the Indian standard IS: 456–2000. Clear cover to the reinforcement bars is specified as 20 mm. Tension reinforcement comprises of 8 mm diameter bars at a centre-line spacing of 150 and 180 mm as shown in Fig. 9.1. The tension reinforcement percentage ensures no flexural failure prior to the shear failure of the specimens. Shear reinforcement is not provided as the slab specimens are designed to fail in shear. Strain gauges are fixed on the steel bars at the mid-span during fabrication to measure steel strain. Concrete strains are measured by strain gauges fixed at d/2 and d distance from the loading block edge; where d is the effective depth of the slab. Punching load is applied by a jack of 30-ton capacity. A steel circular block of 140 mm diameter is placed at the centre of the slab and the jack was used for application of load. The slab specimens are properly levelled and horizontality is maintained prior to testing. Dial gauges are set at the mid-span on the bottom surface exactly below the centre of the loading area to measure mid-span deflection.

The deflection and strains are recorded at a load interval of 0.5 tons. The slab specimens are cured for 28 days and tested at 42 days after white-washing for better visibility of crack growth. All necessary attachments (strain gauges on concrete surfaces) are made prior to testing. The results are recorded in terms of central deflection, punching shear capacity, steel and concrete strain and crack propagation as discussed in following sections.

The two slab specimens for each concrete mix are denoted as the S1 and S2 along with the mix name. The average compressive strength of concrete corresponding to slabs NAC-S1 and NAC-S2 are respectively 49.5 MPa and 52.1 MPa. Similarly, the compressive strength of concrete used in RFA20-S1 and RFA20-S2 slab specimens are 47.3 MPa and 49.0 MPa. RFA30-S1 and RFA30-S2 specimens show compressive strength of 46.4 MPa and 45.0 MPa respectively. It indicates about 5% and 10% reduction in the compressive strength of RFA20 and RFA30 concrete conforming to the respective slab specimens.

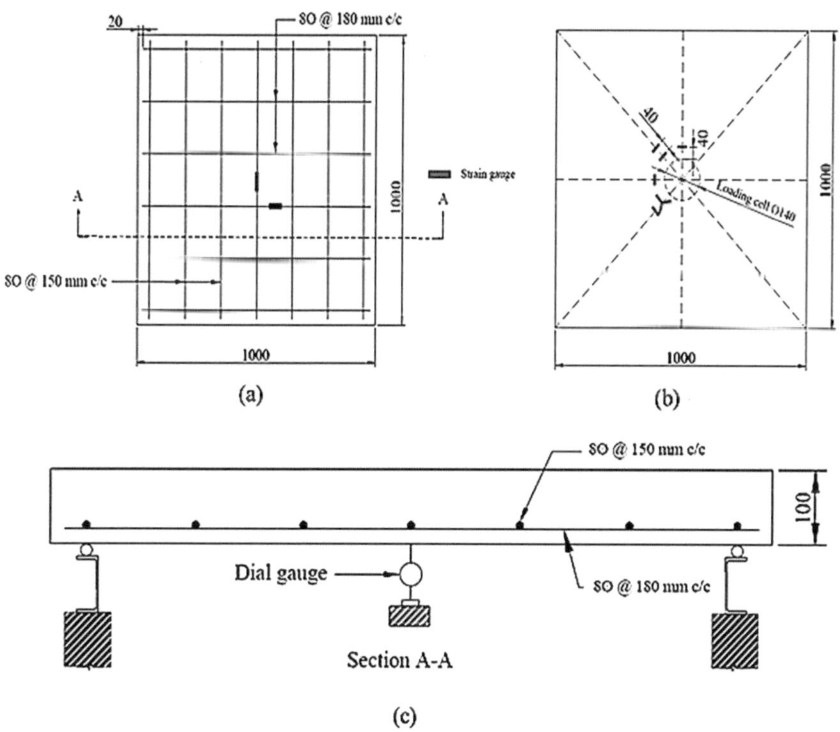

Fig. 9.1 Detailing of RC slabs **a** Plan of slab with reinforcement **b** bottom plan during testing **c** cross-section

9.2.2 Load-Carrying Capacity

Figure 9.2 represents the load–displacement curves of the slab specimens under study. Figure 9.2(a) depicts a typical curve (corresponding to RFA20 specimen) showing response of the specimens under punching load. The specimens behaved linearly up to initiation of cracking, following which stiffness degraded as the cracking progressed until yielding of the main reinforcement. Post-yielding behaviour reflected non-linearity of the load–displacement relation till the ultimate state was reached. Figure 9.2(b) illustrates load–displacement curves of the two similar specimens of each concrete which indicates a slight difference due to the inherent heterogeneity of concrete. The initial change in slope is due to the initiation of crack. The uncracked stiffness of all the specimens was observed to be similar regardless of the types of concrete; but the initial cracking load was slightly higher for NAC (50 kN) than that for RFA30 (40 kN) slabs. About 11% reduction in cracking load was observed for 80% RCA based concrete previously (Reis et al. 2015). The author's earlier studies also reported lowering of cracking strength for reinforced concrete beams comprising FARAC under flexure and shear (Sunayana and Barai 2018, 2020). It may be attributed to the lesser tensile strength (discussed previously in Chap. 6) of coarse RCA based concrete as observed from the material level studies.

After initial cracking, deflection increased at a higher rate due to stiffness degradation; which occurred as a result of crack propagation and the extent of failure (Fig. 9.3). The post-cracking stiffness of the slab specimens was greater for NAC as compared to FARAC. Table 9.1 shows the details of the slab test results. The failure load was 130 kN for NAC and 115 kN for RFA20 as well as RFA30 slab specimens. The incorporation of fly ash in 100% RAC reduced the ultimate load by 11%. Reduction in punching shear strength with increasing RCA replacement percentage was also reported in earlier study (Mahmoud et al. 2018). The ultimate deflection was higher for RFA20 specimens (12.55 mm) as compared to RFA30 even though ultimate load was similar. The compressive strength of concrete on the slab test day was higher in NAC. The normalized punching shear strength was slightly reduced for FARAC; which underlined the predominant effect of the tensile strength of the concrete.

9.2.3 Crack Propagation and Failure Pattern

Figure 9.3 shows the failure mode and crack patterns of the tested slab specimens. In general, radial cracks originated on the bottom surface of the slabs, mostly below the concentrated load and at the outer edge of the loading plate, which further propagated towards the slab edge. The stiffness of the slab specimens started decreasing with formation of many cracks under increasing load. Such stiffness degradation can also be observed from the load–deflection curves discussed in the subsequent section. At the ultimate state, tangential or circumferential cracks were also observed around

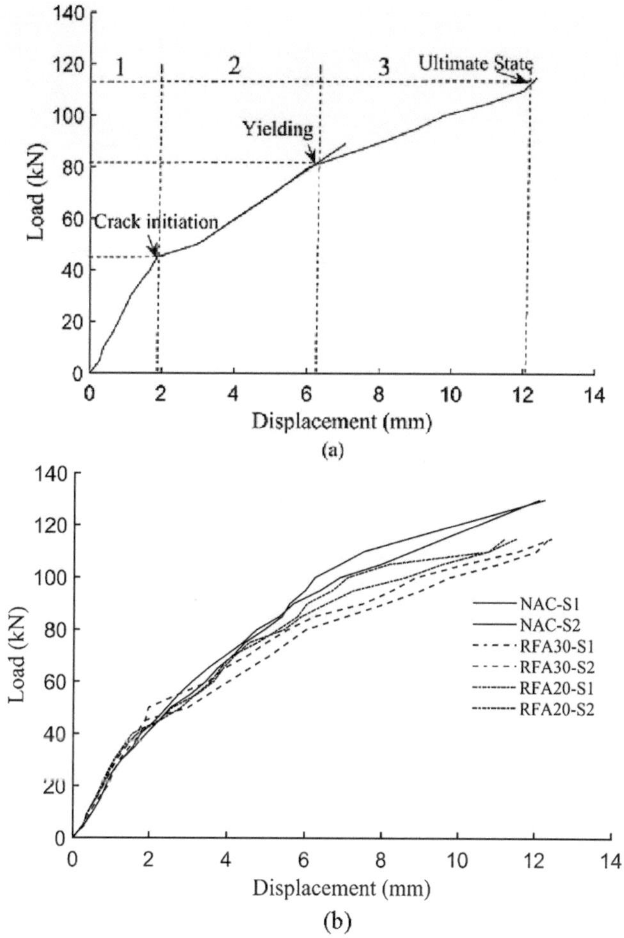

Fig. 9.2 Load–displacement curves of the slab specimens **a** typical curve of RFA30 specimens **b** comparison of different concrete

the punching area which were further connected with the radial cracks. Excessive widening of these cracks leads to the final failure of the specimens.

More prominent radial and tangential shear cracks were observed on the cross-section of FARAC slab specimens. The shear cracks were followed by horizontal cracks below the flexure reinforcement layer which were more prominent in FARAC slabs. This indicates spalling of concrete cover below the reinforcement as the deformation increased rapidly after failure of the slab specimens. Comparatively, a greater extent of cracking, wider cracks and spall-off of the concrete from the bottom surface of specimens were observed in RFA20 and RFA30 as compared to NAC specimens. It may be attributed to lesser tensile strength in the fly ash incorporated recycled

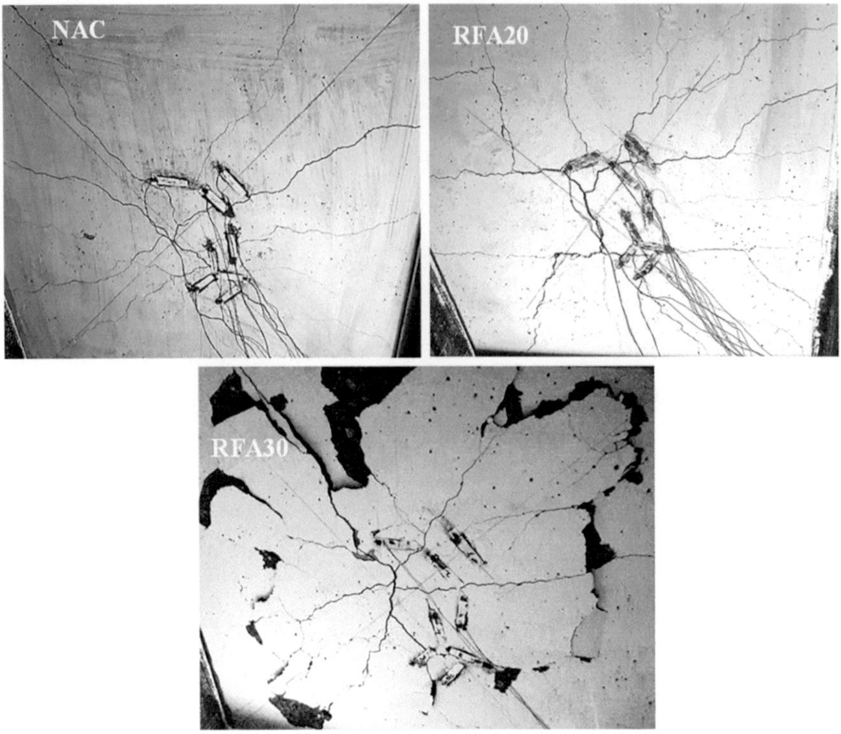

Fig. 9.3 Crack pattern of the slab specimens **a** NAC **b** RFA20 **c** RFA30 bottom face

Table 9.1 Experimental results

Concrete type	Cube strength on slab test day (MPa)	Cracking load (kN)	Ultimate load (kN)	Ultimate deflection (mm)
NAC-S1	50	50	130	12.1
NAC-S2	52	50	130	12.24
RFA20-S1	47	45	115	12.55
RFA20-S2	49	45	115	12.35
RFA30-S1	46	40	115	11.5
RFA30-S2	45	40	115	11.2

aggregate concrete. Cover concrete spall-off might also have been affected by the bond between concrete and reinforcement.

9.2.4 Load-Concrete Strain Relation

Both the radial and tangential cracks are formed with increasing load as discussed earlier. Tangential cracks are associated with the development of tangential strains with increasing load as shown in Fig. 9.4. The rate of increase of strain with load is higher in RFA20 and RFA30 as compared to NAC slab specimens. Ultimate strain in NAC, RFA20 and RFA30 were reported as 0.00106, 0.00126 and 0.00167 respectively. Radial concrete strain (µmm/mm) at the ultimate state were measured as 0.0023, 0.0012 and 0.0005 respectively in NAC, RFA20 and RFA30 slab specimens. These strain values remaining well below the ultimate compressive strain of concrete (0.0035) indicated that the specimens failed in punching rather than concrete crushing.

9.3 Applicability of Existing Design Provisions

The design punching shear of concrete was evaluated as per the provisions of ACI-318, and CEB-FIP model code 2010 following Eqs. 9.1 and 9.2 respectively. As per ACI provisions, the punching shear strength is lesser than or equal to the following three relations.

$$V_c \leq B_{sh} min \left\{ \begin{array}{c} \left(2 + \frac{4}{\beta_c}\right) \frac{\sqrt{f_c}}{12} ud_{av} \\ \left(2 + \frac{\alpha_s d}{u}\right) \frac{\sqrt{f_c}}{12} ud_{av} \\ \frac{\sqrt{f_c}}{3} ud_{av} \end{array} \right\} \tag{9.1}$$

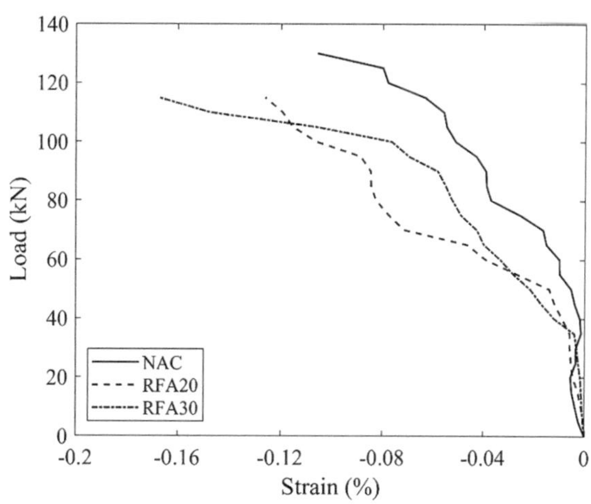

Fig. 9.4 Load vs tangential concrete strain

where d_{av} = average depth of slab (m); β_c = ratio of the long edge to the short edge of the column; α_s = parameter depending on column geometry; u = basic control perimeter at a distance of d/2 from the column edge where d is the distance between extreme compression fibre and centroid of longitudinal reinforcement; and f_c= cylinder compressive strength B_{sh} = shear model factor.

Equation 9.2 shows punching shear strength according to CEB standard based on critical shear crack theory. Here, b_0 is the basic control perimeter at a distance of d/2 from the column edge where d is the distance between extreme compression fibre and centroid of longitudinal reinforcement; k_φ is the parameter depending on rotation of slab and k_{dg} depends on maximum aggregate size d_g.

$$V_c = B_{sh} k_\varphi \sqrt{f_c} b_0 d_v \tag{9.2}$$

Here, $k_\varphi = \frac{1}{1.5+0.9k_{dg}\varphi d}$;$k_{dg} = \frac{32}{16+d_g}$; φ = slab rotation around support area

Key parameters affecting punching shear strength of slab are compressive strength of concrete, maximum aggregate size, effective depth of slab specimens and geometry of column for application of punching load. Dimension, loading or resistance parameters in the actual sense behave as random variables.

However, these factors are considered deterministic in the design equations and the associated variability is taken care of with either a resistance factor or partial safety factor in the design. Moreover, the uncertainty can be appropriately accounted for with a reliability-based approach. Consideration for uncertainty becomes even more significant in the case of recycled aggregate based concrete. The reason is a lack of information about the parent concrete source, moisture content on the surface of RCA, recycling processes. There is an adhered mortar layer on top of RCA, creates two interfaces: one between parent aggregate and old mortar and another between old and fresh mortar. This excess old mortar is distributed unevenly on the RCA surface. Therefore, when applied to RAC, the standard punching shear design specifications for reinforced concrete slabs—which were initially developed for conventional concrete—may not result in a safe design.

The following section aims to determine the failure probability of RAC flat slabs with and without fly ash for limit state of shear. In order to reach the same level of reliability as that of NAC or code-specified target reliability, necessary adjustments are suggested for the safety factors. Reliability of punching shear resistance equations given in the ACI-318 and model code CEB-FIP are assessed.

9.4 Reliability of Existing Design Provisions

The essential steps involved in evaluating reliability are as follows: (i) processing the database to determine the model factor, (ii) identification and statistical characterization of key random variables influencing punching shear resistance, (iii) defining limit states and using Monte-Carlo simulations for evaluation of failure probability and (iv) modifying resistance factors as needed to achieve NAC –equivalent or code-specified target reliability for RAC.

9.4.1 *Processing of Database for Calculation of Model Factor (B_{sh})*

The impact of RCA on the applicability and correctness of shear resistance design models can be represented by a term called the model factor (B_{sh}). B_{sh} is the ratio of shear resistance experimentally measured to that of theoretically predicted accounting for the uncertainties allied with model predictions. Typically, B_{sh} is incorporated as a multiplicative factor in the resistance equation. It is essential to validate the model factors derived from previous experimental data against new datasets to ensure improved prediction reliability of existing resistance models. Notably, variations in material properties driven by changes in raw material production processes and evolving performance standards, can influence the model factors.

The database analysed consisted of shear-critical NAC and RAC slab specimens with no shear reinforcement, all of which failed due to punching shear. The shear resistance models outlined in ACI-318 and the CEB-FIB model code were used to theoretically compute the punching shear capacity as represented in Eqs. 9.1 and 9.2. The method proposed by Holický et al. (2016) was followed for processing the dataset. It involves four key steps: (i) identifying outliers, (ii) verifying the randomness of the data, (iii) fitting an appropriate distribution curve to the dataset, and (iv) calculating the statistical properties of the model factors. Outliers were recognized and eliminated from the database by means of box and whisker plots as shown in Fig. 9.5. The data dispersion was visualized, with vertical lines and the interquartile range (IQR), representing the 25th–75th percentiles, indicating the maximum and minimum values. The data distribution was analysed further and potential outliers were identified as data points lying (i) beyond 1.5 times the IQR or (ii) exceeding 3 times the standard deviation for a given dataset.

The correlation between B_{sh} and other influencing variables was calculated using both the bivariate Pearson and Spearman correlation coefficient. A relatively low correlation supported the randomness of this model factor. Following these above-mentioned procedure, 70 data points were chosen in NAC and 30 datapoints in RAC. This database is summarized in the Table 9.2. The statistical properties such as mean and variance of the respective model factors were calculated based on standard equations. The model factors corresponding to RAC displayed lower mean values and higher coefficient of variation as compared to NAC. This trend suggested that there exists a difference between how these conventional equations apply to RAC and NAC. The model factors were subjected to the Anderson darling test to confirm a normal probability distribution. For the purpose of calculating theoretical shear capacity, the punching shear resistance equations of ACI-318 and the model code CEB-FIP were utilized. Dead load and live load act in the proportion specified in

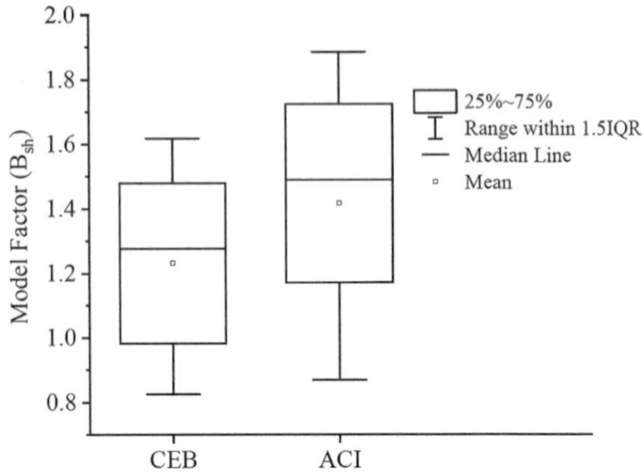

Fig. 9.5 Statistical properties of model factors for the database after removing outliers

the respective standard. It is expected that the material safety factor terms that are applied in these equations are equal to one in order to make accurate predictions regarding actual shear capacity.

The basic variables which affect failure probability of a RC slab under punching shear include: (i) material properties (ii) geometrical parameters and (iii) loads. Live load is assumed to follow extreme type 1 distribution (bias = 1 and COV = 25%) whereas normal distribution is considered for the dead load (bias = 1.05 and COV = 11%). Regarding the statistical characterization of material properties,

Table 9.2 Summary of the database and statistical characterization of model factors

Random Variables	Range of properties for NAC	Range of properties for RAC	Model factors for NAC		Model factors for RAC	
			Mean	COV	Mean	COV
Compressive strength of concrete (MPa) f_c	$20.5 \leq f_c \leq 70$	$29.64 \leq f_c \leq 63.6$	1.30 (ACI) 1.28 (CEB)	0.153 (ACI) 0.157 (CEB)	1.19 (ACI) 1.06 (CEB)	0.198 (ACI) 0.160 (CEB)
Yield strength of steel (MPa) f_y	$255 \leq f_y \leq 752$	$335 \leq f_y \leq 600$				
Effective slab depth (mm) d	$80 \leq d \leq 250$	$50 \leq d \leq 120$				
Mean depth of slab (mm) d_{av}	$64 \leq d_{av} \leq 200$	$50 \leq d_{av} \leq 99$				
Dimension of column (mm) P	$54 \leq P \leq 450$	$100 \leq P \leq 200$				
Control perimeter (mm) u	$576 \leq u \leq 2228$	$667 \leq u \leq 1196$				

Table 9.3 Punching shear resistance factor for RAC

LL/DL	Resistance factor (φ) of RAC	
	ACI	CEB
0.5	0.55	0.56
1	0.56	0.56
1.5	0.57	0.56
2	0.57	0.55
3	0.59	0.56
4	0.59	0.56
Code-specified (φ) for NAC	0.75	0.67

the compressive strength of concrete is considered as a normally distributed random variable. The bias (ratio of mean to nominal values) and coefficient of variation of compressive strength were taken as 1.39 and 8.3% for NAC and 1.31 and 9.2% for RAC respectively (Xiao et al. 2005). Higher COV for compressive strength of RAC is mainly attributed to the randomness in the source of parent concrete as well as adhered mortar content and its distribution on the surface of recycled coarse aggregates. Geometrical properties such as effective depth of the slab and dimension of column are taken as normal distribution with bias unity and COV of 4%.

9.4.2 Reliability Analysis: Results and Discussions

Monte-Carlo simulations are used to generate the statistical characteristics of the fundamental random variables. MATLAB was used to carry out the simulation. Within a single simulation, 50 million sample points were taken into account. The effects of Live load to Dead load ratio (LL/DL) were studied for 0.5, 1, 1.5, 2, 3 and 4 and corresponding design cases were analysed. Probability of punching shear failure was calculated. The load (L) and resistance (R) function can be used to express the shear limit state function (G) as the following: $G(x) = R(x) - L(x)$; where, x stands for a set of random variables. The failure probability was computed for the condition corresponding to $G < 0$. Firstly, the resistance factor recommended for NAC, is used to calculate reliability index (ß) of both NAC as well as RAC (with and without fly ash) slabs. Under comparable design conditions, the resistance factor was recalibrated for RAC provided the ß value of RAC calculated in the first step is lower than that of NAC or the code prescribed target ß value. According to the CEB standard, NAC should achieve a reliability index value 3.8 to ensure a 50-year service life of structure with medium consequence of failure. This recommendation by the CEB standard considers the intended service life of the structure, safety and additional consequences due to injury, loss of life and financial damages etc. The target reliability of RAC slabs was set either as the reliability of NAC calculated in

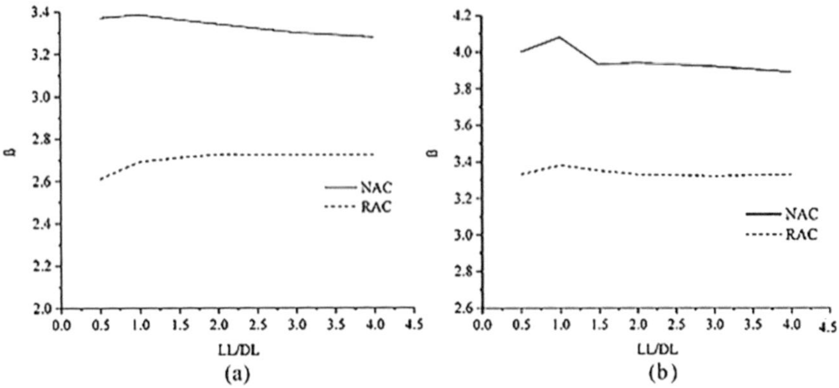

Fig. 9.6 Effect of live-load-to-dead load ratio on punching shear reliability index **a** ACI **b** CEB

the first step or the code specified value, whichever is smaller. The Fig. 9.6 illustrates how the LL/DL ratio affects ß, where ß is calculated following the shear resistance equations of ACI and CEB standard. ß changed slightly up to a LL/DL ratio of 1.5, after which it remained constant regardless of the type of concrete. Higher variability in RAC was indicated by a lower value of ß. The existing shear resistance models show lesser reliability when applied to RAC slabs. Use of same resistance factor in both RAC and NAC showed that RAC slabs failed to meet the target reliability. More precisely, the database that makes up RAC, with or without fly ash, requires calibration of the current standard.

The resistance factors are shown in Table 9.3 after being adjusted for different LL/DL ratio. The reliability analysis indicated that, the shear resistance factors calculated for RAC falls within the range of 0.55–0.59 following ACI standard. Similarly, the resistance factor for punching shear limit state of RAC can be adjusted to 0.55- 0.56 in accordance with CEB standard provisions. Overall, a resistance factor of 0.56 can be employed for the safe and cost-effective design of RAC slabs. It is important to note that this reliability analysis was conducted using a limited database of RC slab specimens composed entirely of recycled coarse aggregates, both with and without fly ash. Future recalibration will be necessary if there are significant variations in the statistical properties of the model factors, caused by new addition to the database.

9.5 Summary

- The role of coarse RCA on the punching shear behaviour of slabs are not very clear from the existing studies. Several studies indicated a decrease in ultimate punching load by 15–20% as coarse RCA content increased. Conversely, no substantial effect of RCA was noted in certain instances.

- Fly ash incorporated RAC slab specimens exhibited marginally reduced cracking load and higher degradation of stiffness after cracking. The ultimate punching load was reduced by about 11% in FARAC as compared to the punching strength of natural aggregate concrete slabs.
- The failure of FARAC specimens was associated with increased spalling of the cover concrete, leading to a more extensive damage. It was also characterized by formation of numerous radial cracks. In FARAC specimens, cracking initiated at a lower load. Higher tensile stress was developed at the steel–concrete interface, which leads to failure of the cover concrete. Such failure was caused due to the lower tensile strength of FARAC.
- The structural safety of FARAC flat slabs with respect to the punching shear loading is evaluated by performing reliability analysis. The shear resistance provisions of ACI and CEB standard for natural aggregate concrete have been recalibrated for RAC, with modifications made to the shear resistance factor.
- Statistical analysis of the database of RAC flat slabs revealed greater variability in the shear model factors when computed using the ACI and CEB shear resistance equations. Additionally, there was also a greater degree of variation in the compressive strength of RAC compared to NAC. As a consequence, RAC flat slabs that were designed with the same punching shear resistance factor as that was prescribed for NAC ended up having a lower reliability index.
- Based on the reliability analysis study discussed in this chapter, a resistance factor of 0.56 was recommended through recalibration of the shear capacity equations for the safe and cost-effective design of RAC slabs. Nevertheless, experimental database can be expanded in future to guarantee greater applicability of this suggested resistance factor.
- Future research in this area is encouraged to investigate the punching shear behaviour of reinforced concrete RAC slabs in the presence of higher fly ash percentage as well as other binary and ternary blended cement incorporated into RAC. The application of the modified resistance factors can be further generalized with the availability of additional data points used for their estimation.

References

ACI 318–11 (2011) Building code requirements for structural concrete and commentary, American Concrete Institute, Farmington Hills, Michigan, USA

FIB Model code for concrete structures (2010) Ernst and Sohn

Francesconi L, Pani L, Stochino F (2016) Punching shear strength of reinforced recycled concrete slabs. Constr Build Mater 127:248–263

Holický M, Retief JV, Sýkora M (2016) Assessment of model uncertainties for structural resistance. Probab Eng Mech 45:188–197

IS (Indian Standard) (2000) Plain and reinforced concrete code of practice. IS: 456–2000, Bureau of Indian Standards, New Delhi, India

Mahmoud ZI, Saeed KS (2018) Punching shear behavior of recycled aggregate reinforced concrete slabs. Alex Eng J 57(2):841–849

Reis N, de Brito J, Correia JR, Arruda MR (2015) Punching behaviour of concrete slabs incorporating coarse recycled concrete aggregates. Eng Struct 100:238–248
Sacramento PVP, Ferreira MP, Oliveira DRC, Melo GS (2012) Punching strength of reinforced concrete flat slabs without shear reinforcement. Revista IBRACON De Estruturas e Materiais 5:659–691
Sahoo S, Singh B (2020) Recycled aggregate concrete slab punching shear capacity. Structures 24:426–443
Sunayana S, Barai SV (2018) Flexural performance and tension-stiffening evaluation of reinforced concrete beam incorporating recycled aggregate and fly ash. Constr Build Mater 174:210–223
Sunayana S, Barai SV (2020) Shear behavior of fly-ash-incorporated recycled aggregate concrete beams. ACI Struct J 117(1):289–303
Xiao J, Li J, Zhang C (2005) Mechanical properties of recycled aggregate concrete under uniaxial loading. Cem Concr Res 35(6):1187–1194
Xiao J, Wang W, Zhou Z, Tawana MM (2018) Punching shear behaviour of recycled aggregate concrete slabs with and without steel fibres. Front Struct Civ Eng 13:725–740

Part IV
Durability and Long-Term Concrete Properties

Chapter 10
Durability and Long-Term Properties of Concrete with SCMs and Coarse RCA

Preview

One of the most essential aspects for ensuring durability as well as sustainability of concrete lies in proper selection of construction materials and design which ensure a longer service life of the structure. Porosity and connectivity of pores essentially regulate the durability of concrete. As discussed before, the attached mortar layer in coarse RCA is porous, less dense and contains microcracks; this inherent inferiority poses a major concern for making durable concrete. This chapter focuses on durability, shrinkage, and creep properties of concrete when coarse RCA and SCMs are incorporated. Durability properties of concrete has been assessed in terms of permeability, chloride ion penetration, sulphate attack and carbonation etc.

10.1 Permeability

10.1.1 Role of Coarse RCA

RAC shows higher water absorption which increases further with replacement percentage of coarse RCA due to greater water absorption of aggregates, higher porosity, and pore connectivity. In the same way, permeability increases due to higher RCA content in concrete but decreases at longer curing age on account of greater extent of hydration (Kwan et al. 2012). This effect of higher permeability in RAC owing to microcracks in the adhered mortar layer was also observed in other studies (Bravo et al. 2015; de Brito et al. 2016; Guo et al. 2018). 100% coarse RCA causes an increase in water absorption by 2.47 times (water absorbed by immersion) and 1.7 times (by capillary action). 100% fine RCA shows 46% higher absorption than NAC due to the presence of more capillary channels (Evangelista and Brito 2010). Other factors such as size of RCA, crushing method, and parent concrete source etc.

S. Sunayana and S. V. Barai, *Green Concrete*, Springer Transactions in Civil and Environmental Engineering, https://doi.org/10.1007/978-981-96-6982-0_10

(Guo et al. 2018) are also found to have influence on concrete permeability. Larger the particle size of RCA, lesser the tortuosity of flow path and bleeding effect which attribute towards increasing permeability of gases.

10.1.2 Role of SCMs in RAC

The two most important role of SCMs in concrete are (1) filler effect due to finer particle sizes and (2) pozzolanic reactivity in an alkaline environment. Both these effects contribute towards reducing porosity and densifying the microstructure of concrete. This suggests a potential reduction of permeability of RAC in the presence of SCMs. Fly ash used as cement or fine aggregate replacement in RAC lower water permeability due to pore-refinement and this effect is more significant for higher replacement of recycled aggregates (Faella et al. 2016). Similarly, Somna et al. (2012) reported that 20% ground fly ash as cement substitute can lower water permeability of RAC due to finer particles filling the voids in the microstructure. With regards to water permeability, the positive effects of fly ash in RAC are more prominent due to pozzolanic reactions and reduced interconnectivity of pores at longer curing age (90 days and beyond) as compared to that at 28 days (Kurda et al. 2019a). They reported that the water absorption by immersion reduced by 3 and 15% for 30 and 60% fly ash replaced RAC at 28 days. The observed effect was ascribed to a reduced water requirement in the presence of fly ash when creating a workable mix, consequently enhancing the density of the paste. The water requirement in fly ash incorporated concrete is also affected by the shape and size of fly ash particles. High-volume fly ash (60%) and 100% RCA along with superplasticizer can be used for a reduced permeability; superplasticizer helps in better dispersion of cementitious particles and hence increases reactivity (Kurda et al. 2019a). The role of GGBFS (40 and 60%) in reducing water absorption was reported due to reduction in pore connectivity and densification of microstructure (Majhi and Nayak 2019). Ultrafine cementitious materials such as silica fume and metakaolin show positive effects in terms of improving impermeability of RAC. This effect is attributed to the microcrystalline nucleation effects of finer materials, which enhance the hydration reactions (Guo et al. 2018).

10.2 Chloride Attack

Chloride attack in reinforced concrete structure can lead to corrosion of reinforcement by destroying the passive protective layer; hence, negatively affects durability of reinforced concrete structures. Transport of chloride ions into concrete is a complex phenomenon involving water diffusion and capillary absorption. This is regulated by accessible porosity and pore connectivity (Fenaux et al. 2019).

10.2.1 Role of Coarse RCA

Use of coarse RCA in concrete increase the chloride permeability which is further enhanced by the replacement percentage of RCA; higher the RCA content greater the chloride penetration (Bravo et al. 2015; Faella et al. 2016). For example, total charge passed in RAC with 100% coarse RCA is 2.07 times that of NAC. The weak and porous adhered mortar layer and higher water absorption in coarse RCA led to formation of more transport channels for chloride penetration (Zhang and Bao 2023). The chloride ions that penetrate the concrete react with $Ca(OH)_2$ in the vicinity of the old and new interfacial transition zones (ITZs), leading to the formation of expansive salts. This process contributes to the creation of additional cracks in RAC (Liu et al. 2020). They also suggested that the old mortar layer, being more porous with larger pores (diameter or volume), microcracks and connected channels, facilitates faster diffusion of chloride ions and extra space for crystallization of salts. Other factors influencing the chloride permeability of concrete may be cited as the strength of parent concrete and amount of adhered mortar content (Kou and Poon 2015; Duan and Poon 2014). Higher strength of parent concrete and lesser old mortar ensure improved chloride resistance.

10.2.2 Role of SCMs in RAC

Addition of mineral admixtures as cement substitute in RAC display positive effects in terms of reducing chloride permeability. This observation is attributed to improvement of the microstructure due to pore refinement effect as well as secondary hydration. Fly ash (at 25 and 35%) used as cement replacement or cement addition in RAC increase the chloride ion penetration resistance (Kou and Poon, 2012). Ground fly ash used as 20, 35 and 50% in RAC also shows similar effects of lowering the chloride penetration depth compared to RAC without fly ash (Somna et al. 2012). The reasons for this phenomenon were ascribed to the ability of fine fly ash particles in reducing the average pore size and blocking chloride ions from passing through the C–S–H generated during hydration and pozzolanic reactions. Binary or ternary blended cement such as 20% fly ash with 10% silica fume or metakaolin enhance the chloride resistance of RAC at longer curing ages of 90 or 120 days (Kapoor et al. 2016). Use of 30% or 60% GGBFS in 60% coarse RCA based concrete reduced the chloride permeability by 50% and 67% respectively. This is due to enhanced homogeneity caused by reduction in pore connectivity in the microstructure and early pozzolanic reactions (Hu et al. 2019). The effect of silica fume on chloride permeability resistance of RAC was examined by Sasanipour et al. (2019). They observed that, total charge passed in a chloride permeability test increases with increasing coarse RCA replacement (18% higher at 100% coarse RCA) indicating higher permeability due to porous adhered mortar. However, 8% silica fume can reduce the chloride penetration by 28% in 100% coarse RCA based concrete. These positive effects of silica fume were caused due to improvement in pore network and formation of high-quality C–S–H in the presence of fine silica fume particles.

10.3 Sulphate Attack

Concrete subjected to aggressive environment may be prone to sulphate attack which is one of the major durability concerns in concrete. Sulphate attack deteriorate concrete on account of two phenomena as follows: (1) externally present sulphate ions percolate into the concrete surface at contact and react with hydration products like $Ca(OH)_2$ resulting harmful compounds like ettringite and calcium sulphate. These products cause volumetric expansion of concrete there by developing internal stresses and cracks (Santhanam et al. 2002). This chemical reaction also causes decalcification of C–S–H and may weaken the hardened paste (Massaad et al. 2016). (2) Secondly, the physical process where the sulphate solution, upon evaporation, results in salt crystallization near the surface concrete. This salt crystallization may lead to cracking of concrete if the crystallization pressure exceeds the tensile strength of concrete (Irassar et al. 1996).

10.3.1 Role of Coarse RCA

The effect of RCA (replacement percentage of 30, 50,70, and 100) on sulphate resistance of concrete examined by Qi et al. (2017) highlighted an improved performance till 50% RCA due to higher water absorption and reduced effective water-binder ratio. However, at higher replacement level, there was a reduction in sulphate resistance; cracks developed in the concrete due to expansion of the reaction products. In another recent study, the role of RCA on sulphate resistance of concrete was studied by Zhang et al. (2020a, b, c) using 100% replacement of coarse aggregates and two types of sulphate environment (full-soaking and dry–wet cycles). They reported higher sulphate ions uptake in RAC than NAC at a fixed depth from the exposure surface of the specimen. Such behaviour was attributed to the presence of higher microcracks in the old adhered mortar layer which led to greater sulphate ion transport and porous framework causes more accumulation as compared to NAC. The dry–wet cycle environment accelerates the SO_4^{2-} transport as compared to full-soaking. Early occurrences and faster growth of surface scaling, crack width and concrete spalling were noticed in RAC due to weaker ITZs and inferior mechanical properties.

10.3.2 Role of SCMs in RAC

Pozzolanic materials show positive effects in reducing the level of damage due to sulphate attack; factors responsible may be finer particle size, improvement in microstructure due to the pozzolanic reactions, reduced ion diffusivity and lesser availability of C_3A. Class F fly ash is more effective (35% reduction of expansion as compared to control concrete) in resisting sulphate attack in concrete as compared to class C fly ash due to low C_3A content of the earlier (Bonakder and Mobasher 2010).

The application of SCMs in RAC showed beneficial effects in terms of sulphate resistance as compared to RAC without it. Qi et al. (2017) considered mass change, on immersing the concrete specimens in sodium sulphate solution, as a parameter to indicate degradation under sulphate attack. The maximum mass change percentage after 270 days were observed to be 2.4, 1.07 and 1.01% for specimens with 100% RCA without any SCMs, with 30% fly ash and 30% granulated blast furnace slag (GBFS) in 100% RCA based concrete respectively. The increased sulphate resistance in the presence of fly ash and GBFS were attributed to the refinement of pores, higher pozzolanic reaction implying greater CH consumption and lesser CH availability to form gypsum (lower gypsum and ettringite peaks were evident from XRD analysis). Gao et al. (2024) studied the effects of 8, 16 and 24% fly ash as mineral admixture and reported an increased sulphate resistance due to pozzolanic reactions and formation of higher C–S–H gels. $SO_4{}^{2-}$ ion content reduced to a minimum in the presence of 24% fly ash in RAC.

10.4 Carbonation

Carbonation occurs, when CO_2 diffuse into concrete from any external sources, in the presence of moisture forms carbonic acid which further reacts with Ca $(OH)_2$ to form $CaCO_3$. This reaction process can be represented in terms of Eqs. 10.1–10.3 (Kurda et al. 2019b). This carbonates, though beneficial for reducing permeability, reduces pH of concrete and destroys the protective coating on reinforcement (Silva et al. 2015a). Thus, the process of carbonation also leads to increasing corrosion of reinforcement similar to the impact of chloride ion ingress as discussed before.

$$H_2O + CO_2 \rightarrow HCO_3 + H^+ \quad (10.1)$$

$$HCO_3^- \rightarrow H^+ + CO_3^{2-} \quad (10.2)$$

$$Ca(OH)_2 + 2H^+ + CO_3^{2-} \rightarrow CaCO_3 + 2H_2O \quad (10.3)$$

10.4.1 Role of Coarse RCA

There are different factors which might influence the carbonation resistance of recycled aggregate concrete such as percentage replacement, crushing procedures, quality of RCA, and curing conditions, mineral admixtures and use of other water reducing admixtures etc. Increase of coarse RCA replacement in concrete results in lower carbonation resistance due to higher porosity of the attached mortar layer (Gomes and de Brito 2009). Carbonation depth of concrete with 100% coarse RCA increases by about 2.5 times as compared to NAC (Silva et al. 2015a). On the contrary, few studies showed equivalent or greater carbonation resistance of RAC due to higher water absorption of RCA (air-dried aggregates) resulting lower effective w/c ratio (Ridzuan et al. 2005) or using equivalent mortar volume method of mix design (Fathifazl et al. 2009). Coarse RCA produced using both primary and secondary crushing stage, when used in concrete, showed lower carbonation depth (Pedro et al. 2014) as compared to the RCA processed by only primary crushing. The reasons for this improved carbon resistance in the earlier case were identified as (i) improved quality of coarse RCA with less adhered mortar content (ii) generation of more spherical shape aggregates which improve packing of the concrete and (iii) less water requirement for same workability level in spherical aggregates reduces porosity. RCA produced from high strength parent concrete shows improved carbonation resistance. For 50 MPa and 20 MPa parent concrete compressive strength, the carbonation depth increased by 10% and 80% respectively with respect to the control NAC (Xiao et al. 2012). On the other hand, few studies concluded that the attached mortar content and source concrete strength showed little influence on carbonation resistance of RAC relative to that of NAC, provided the water absorption values of coarse RCAs are similar (Dosho 2007; Eguchi et al. 2007). Curing conditions may affect the carbonation resistance of RAC in the way that concrete cured in the dry environment (relative humidity 65%) showed higher depth of carbonation which increases further with increasing replacement percentage of coarse RCA (Amorim et al. 2012). Using water reducing admixtures may be beneficial in this context of carbonation resistance due to two main reasons (i) higher internal humidity can be maintained even in the presence of low w/c ratio which would lower the water evaporation from within the concrete and (ii) superplasticizer may delay the hydration effects. Both these effects impacted RAC as if the curing effects were prolonged hence, reduce depth of carbonation (Matias et al. 2014).

10.4.2 Role of SCMs in RAC

Concrete containing SCMs experiences two parallel processes when CO_2 enters inside it, the performance of concrete is regulated by the dominated process (Neville 1995). Firstly, presence of lower Ca $(OH)_2$ content in the blended cement facilitates greater carbonation since less CO_2 is required for the reaction. Secondly, higher pozzolanic reactions could result in a denser microstructure with more C–S–H thereby decreasing pores and diffusivity of CO_2 particles; thus, carbonation may get reduced.

The effects of SCMs on the carbonation resistance of RAC showed contradictory results in different studies as discussed below. In RAC with fly ash, Kou and Poon (2013) reported increased carbonation due to lesser calcium hydroxide content even after 10 years (90% higher in presence of 100% coarse RCA and 55% fly ash). However, RAC incorporating fly ash showed greater carbonation resistance when lower w/b ratio and higher total cementitious materials were considered for the mix design due to reduced permeability and higher $Ca(OH)_2$ content (Corinaldesi and Moriconi 2009). Xiao et al. (2012) investigated the effect of 10% fly ash, silica fume and slag on the carbonation behaviour of RAC. They reported reduction of porosity and internal pore structures due to these SCMs, however, depth of carbonation still increased due to lower alkali content in the blended cement. The key factors having an influence on carbonation resistance of fly ash incorporated concrete may be listed as water-to-binder ratio (w/b), fly ash content, relative humidity (RH), environmental temperature, and porosity of concrete (Kurda et al. 2019b). For concrete with 100% coarse RCA and 30% fly ash, depth of carbonation was observed to be almost 2 to 4 times after 28 days and 365 days respectively. Higher w/b ratio leads to increased porosity and hence permeability thereby increasing the depth of carbonation (da Silva and Oliveira Andrade 2017). As H_2O is required for the carbonation reaction, a relative humidity of 50% was found to be optimum (Papadakis et al. 1991; Limbachiya et al. 2012) when compared with the dry condition (RH < 40%) and wet state (RH > 90%). Slag and Metakaolin were found to improve carbon resistance because finer SCMs particles cause improvement in the microstructure with lesser pores (Sun et al. 2006; Singh and Singh 2016).

10.5 Shrinkage

Shrinkage phenomenon have an impact on performance of concrete and hence regulates structural design for service limit state. More shrinkage deformation could lead to formation of cracks and jeopardise the structural safety (Mao et al. 2021). Aggregates in concrete provide restraints to shrinkage of the paste. This shows the significance of discussing shrinkage (autogenous and drying) in this context where recycled coarse aggregates are used instead of the conventional aggregates to make concrete.

10.5.1 Role of Coarse RCA

Autogenous shrinkage constitutes a smaller part of the total shrinkage of concrete and it is critical in the case of high strength or high performance concrete due to low w/b ratio (Jensen and Lura 2006). Autogenous shrinkage of RAC mainly depends on the internal curing effect of RCA owing to the higher porosity and water absorption of these aggregates (Zhang et al. 2020a, b, c); higher the effect of internal curing, lower is the autogenous shrinkage. This internal curing effect is lower in coarse RCA as compared to the light-weight aggregates (Seara-Paz et al. 2016); still, it is significant in lowering the autogenous shrinkage of RAC that is greater reduction (18–80%) at higher percentage replacement (Zhang et al. 2020a; Mao et al. 2021). Other factors such as saturation degree of RCA, strength of parent concrete and particle size might influence the internal curing effect of coarse RCA. 100% saturated coarse RCA can reduce the autogenous shrinkage by 35% as compared to the RCA in dry state (Yildrim et al. 2015). Lower parent concrete strength resulted in coarse RCA with greater water absorption and porosity; hence helps in reducing autogenous shrinkage (Gonzalez-Corominas and Etxeberria 2016).

In general, higher porosity and smaller size pores (less than 50 nm) influence the drying shrinkage of concrete adversely (Mehta and Monteiro 2006). As mentioned before, coarse RCA have greater porosity in the adhered mortar layer (unless the old mortar layer is treated). Due to higher porosity as well as excess total paste content (due to old mortar), concrete incorporating coarse RCA are inherently prone to higher drying shrinkage (He 2023). Drying shrinkage of concrete comprising coarse RCA depends on the properties of aggregates, particle size, quality, and quantity of adhered mortar content, mix design, and mixing (Mao et al. 2021). Domingo et al. (2010) reported about 80% increase in the drying shrinkage when 100% coarse RCA is used in concrete due to higher porosity. Higher porosity of the attached mortar layer is the reason for higher porosity of RAC, this gets reflected in the higher drying shrinkage (Duan and Poon 2014). In general, besides the total porosity, pore-size distribution indicating more number of smaller size pores (<50 nm) plays crucial role in increasing drying shrinkage of concrete (Meheta and Monteiro 2006). Coarse RCA cause more drying shrinkage in concrete as compared to fine RCA; due to larger particle size and higher volume fraction of coarse aggregates (55% or more) as compared to fine aggregates (about 27%). Application of 100% coarse RCA causes an increase in drying shrinkage of concrete by 21–77% whereas these values range from 17 to 62% in the case of 100% fine RCA (Corinaldesi 2010; Zhang et al. 2020b). The restraint to shrinkage provided by coarse RCA may be lower on account of its smaller elastic modulus (Silva et al. 2015b).

10.5.2 Role of SCMs in RAC

SCMs used as a partial replacement of fine aggregate in RAC was found to have negative effect with regards to the drying shrinkage of concrete. For example, as compared to NAC, an increase of about 34% drying shrinkage was reported in RAC with 50% fine RCA replaced fly ash (Corinaldesi and Moriconi 2009) due to a reduction in elastic modulus. Use of fly ash as cement replacement or additions at 25 and 35% in RAC reduce the drying shrinkage due to the dilution effect (Kou

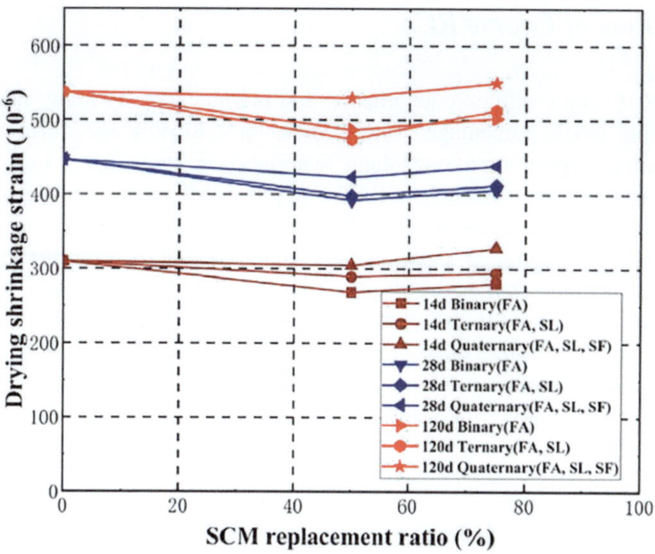

Fig. 10.1 Effects of binary and ternary blended cements on drying shrinkage of RAC (Adopted from Mao et al. 2021)

and Poon 2012). This dilution effect indicates a reduction in effective water-cement ratio with increasing cementitious materials (Atis et al. 2004). The use of high SCMs content (50 and 75%) to make binary (cement with fly ash), ternary (cement with fly ash and GBFS) or quaternary (cement with fly ash, GBFS and silica fume) mixes are beneficial in reducing the drying shrinkage strain of RAC due to lowering of cement content and consequently decreasing heat of hydration in the pozzolanic mixes (Guo et al. 2020). The drying shrinkage strain of binary and ternary blended cement used in RAC is adopted from Mao et al. (2021) as illustrated in Fig. 10.1. It was observed from the figure that, with increase of SCM replacement until 50%, drying shrinkage strain reduces. However, drying shrinkage strain increases for higher SCM replacement percentage.

10.6 Creep

Creep is the phenomenon of increasing deformation or strain of any structural elements at constant loading. Both the phenomena of creep and drying shrinkage are initiated in the hydrated cement paste (HCP); primarily caused by the elimination of physically adsorbed water from the HCP (Mehta and Monteiro 2006). As discussed before, the loss of adsorbed water due to relative humidity difference between the concrete and outside environment cause shrinkage; whereas water loss due to magnitude and duration of applied sustained stress cause creep.

10.6.1 Role of Coarse RCA

Creep of RAC is a complex phenomenon and is influenced by many factors such as RCA replacement percentage, water-cement ratio, loading age, curing duration, porosity, and incorporation of pozzolanic materials (Lye et al. 2016). There state-of-the-art report suggested that creep of RAC is adversely affected due to higher adhered mortar content and porosity, weaker ITZ, lower elastic modulus, greater cement requirement for satisfying strength criteria, and increased total mortar content etc. Also, based on creep studies in different countries (Belgium, Holland, Spain, and Switzerland) conducted till 2011, a relative increase of 25%, 25–45%, 20–60% and 30–40% creep was reported respectively in 100% coarse RCA based concrete. Creep of RAC increases linearly with increasing the replacement level of RCA (Silva et al. 2015a, b, c). For example: for 100% coarse RCA, creep increases by 60%.

10.6.2 Role of SCMs in RAC

SCMs such as fly ash as partial cement replacement in RAC was observed to have beneficial effects on creep of concrete (Kou et al. 2007). For example: 25% cement substituted fly ash in RAC with coarse RCA replacement of 20%, 50% and 100% is effective in decreasing creep by 13–16% (Kou and Poon 2012). Similarly, the creep of RAC (with 20, 50 and 100% coarse RCA) reduced by 19–20% when 35% fly ash was used as cement replacement. These positive effects of fly ash on creep properties of concrete were attributed to lower stress/strength ratio due to pozzolanic reactions at longer curing ages (He, 2023). The spherical shape and finer particle size distribution of fly ash and silica fume are supposed to be effective positively towards creep reduction in RAC due to improvement in packing of the paste and long-term strength gain due to pozzolanic reactions (Lye et al. 2016). Effect of fly ash and coarse RCA on the creep strain of concrete is illustrated in Fig. 10.2. It was observed from the figure that creep strain increases with increase in coarse RCA replacement percentage. Fly ash, when used in addition to cement in RAC, is more effective in reducing creep.

10.7 Summary

- The density of concrete constituted of recycled concrete aggregates is lower due to the porous and weak old mortar. This is reflected in higher tendency of water absorption, chloride permeability, lower stiffness, higher carbonation, corrosion and negatively impacted long-term durability properties. Lower stiffness of coarse RCA leads to higher shrinkage.

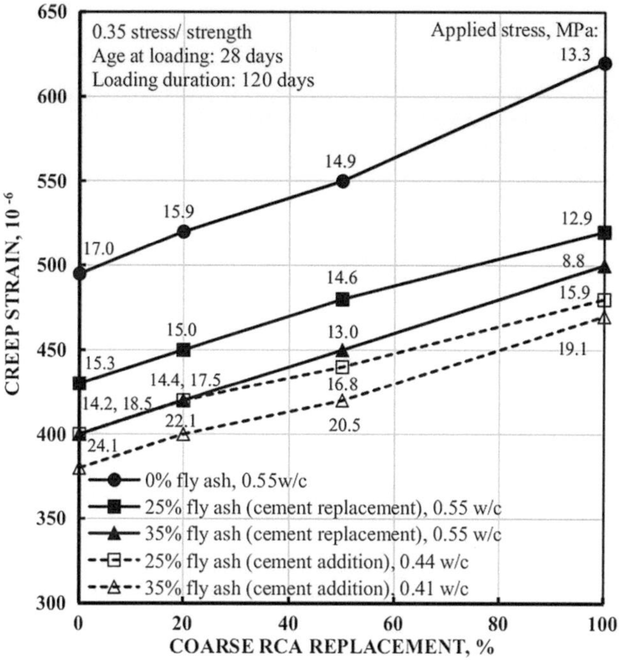

Fig. 10.2 Effect of coarse RCA and fly ash on creep strain of concrete (Adapted from Lye et al. 2016)

- The durability of RAC can be improved by strengthening the old mortar layer, reducing this adhered mortar content, and adopting a suitable mix design. The removal or strengthening of adhered mortar methods available for the treatment of RCA are effective but are cost and time intensive; thereby limiting their practical application. Further research is needed in this regard.
- The durability properties of RAC can be improved by incorporating SCMs. Using fly ash in RAC, chloride ion penetration, carbonation, shrinkage, and creep reduces. Slag as a partial cement substitute can be beneficial in reducing chloride and sulphate attacks in RAC; but ineffective towards shrinkage and water absorption. Silica fume can reduce water absorption and chloride penetration of RAC.
- Blended SCMs such as fly ash in combination with slags or combined fly ash, slag and silica fume can also be effectively utilized for enhancing the durability properties of RAC.

References

Amorim P, De Brito J, Evangelista L (2012) Concrete made with coarse concrete aggregate: influence of curing on durability. ACI Mater J 109(2):195–204

Atis CD, Kilic A, Sevim UK (2004) Strength and shrinkage properties of mortar containing a nonstandard high-calcium fly ash. Cem Concr Res 34(1):99–102

Bonakdar A, Mobasher B (2010) Multi-parameter study of external sulfate attack in blended cement materials. Constr Build Mater 24(1):61–70

Bravo M, de Brito J, Pontes J, Evangelista L (2015) Durability performance of concrete with recycled aggregates from construction and demolition waste plants. Constr Build Mater 77:357–369

Corinaldesi V (2010) Mechanical and elastic behaviour of concretes made of recycled-concrete coarse aggregates. Constr Build Mater 24(9):1616–1620

Corinaldesi V, Moriconi G (2009) Influence of mineral additions on the performance of 100% recycled aggregate concrete. Constr Build Mater 23(8):2869–2876

da Silva SR, de Oliveira Andrade JJ (2017) Investigation of mechanical properties and carbonation of concretes with construction and demolition waste and fly ash. Constr Build Mater 153:704–715

de Brito J, Ferreira J, Pacheco J, Soares D, Guerreiro M (2016) Structural, material, mechanical and durability properties and behaviour of recycled aggregates concrete. J Build Eng 6:1–16

Domingo A, Lázaro C, Gayarre FL, Serrano MA, López-Colina C (2010) Long term deformations by creep and shrinkage in recycled aggregate concrete. Mater Struct 43:1147–1160

Dosho Y (2007) Development of a sustainable concrete waste recycling system-application of recycled aggregate concrete produced by aggregate replacing method. J Adv Concr Technol 5(1):27–42

Duan ZH, Poon CS (2014) Properties of recycled aggregate concrete made with recycled aggregates with different amounts of old adhered mortars. Mater des 58:19–29

Eguchi K, Teranishi K, Nakagome A, Kishimoto H, Shinozaki K, Narikawa M (2007) Application of recycled coarse aggregate by mixture to concrete construction. Constr Build Mater 21(7):1542–1551

Evangelista L, de Brito J (2010) Durability performance of concrete made with fine recycled concrete aggregates. Cement Concr Compos 32:9–14

Faella C, Lima C, Martinelli E, Pepe M, Realfonzo R (2016) Mechanical and durability performance of sustainable structural concretes: an experimental study. Cement Concr Compos 71:85–96

Fathifazl G, Abbas A, Razaqpur AG, Isgor OB, Fournier B, Foo S (2009) New mixture proportioning method for concrete made with coarse recycled concrete aggregate. J Mater Civ Eng 21(10):601–611

Fenaux M, Reyes E, Gálvez JC, Moragues A (2019) Modelling the transport of chloride and other ions in cement-based materials. Cement Concr Compos 97:33–42

Gao S, Qin Z, Long W, Guo X, Ji Y, Zhang H, Xing F, Wang S, Liu A (2024) The sulfate transfer characteristics in recycled aggregate concrete incorporated with fly ash under percolation theory. Case Stud Constr Mater 20:e02774

Gomes M, de Brito J (2009) Structural concrete with incorporation of coarse recycled concrete and ceramic aggregates: durability performance. Mater Struct 42:663–675

Gonzalez-Corominas A, Etxeberria M (2016) Effects of using recycled concrete aggregates on the shrinkage of high-performance concrete. Constr Build Mater 115:32–41

Guo H, Shi C, Guan X, Zhu J, Ding Y, Ling TC, Zhang H, Wang Y (2018) Durability of recycled aggregate concrete–a review. Cement Concr Compos 89:251–259

Guo Z, Jiang T, Zhang J, Kong X, Chen C, Lehman DE (2020) Mechanical and durability properties of sustainable self-compacting concrete with recycled concrete aggregate and fly ash, slag and silica fume. Constr Build Mater 231:117115

He Z (2023) Creep and shrinkage performance of high-performance recycled aggregate concrete. In: Multi-functional concrete with recycled aggregates. Woodhead Publishing, pp 145–177

Hu W, Li S, Song C, Chen Z, Chen L, Yang Y, Luo W (2019) A laboratory analysis of chloride ions penetration in recycled aggregates concrete admixed with ground granulated blast furnace slag.

In: IOP conference series: materials science and engineering, vol 611, no. 1. IOP Publishing, p 012051
Irassar EF, Di Maio A, Batic OR (1996) Sulfate attack on concrete with mineral admixtures. Cem Concr Res 26(1):113–123
Jensen OM, Lura P (2006) Techniques and materials for internal water curing of concrete. Mater Struct 39(9):817–825
Kapoor K, Singh SP, Singh B (2016) Durability of self-compacting concrete made with recycled concrete aggregates and mineral admixtures. Constr Build Mater 128:67–76
Kou SC, Poon CS (2012) Enhancing the durability properties of concrete prepared with coarse recycled aggregate. Constr Build Mater 35:69–76
Kou SC, Poon CS (2013) Long-term mechanical and durability properties of recycled aggregate concrete prepared with the incorporation of fly ash. Cement Concr Compos 37:12–19
Kou SC, Poon CS (2015) Effect of the quality of parent concrete on the properties of high performance recycled aggregate concrete. Constr Build Mater 77:501–508
Kou SC, Poon CS, Chan D (2007) Influence of fly ash as cement replacement on the properties of recycled aggregate concrete. J Mater Civ Eng 19(9):709–717
Kurda R, de Brito J, Silvestre JD (2019a) Water absorption and electrical resistivity of concrete with recycled concrete aggregates and fly ash. Cement Concr Compos 95:169–182
Kurda R, de Brito J, Silvestre JD (2019b) Carbonation of concrete made with high amount of fly ash and recycled concrete aggregates for utilization of CO_2. J CO_2 Utilization 29:12–19
Kwan WH, Ramli M, Kam KJ, Sulieman MZ (2012) Influence of the amount of recycled coarse aggregate in concrete design and durability properties. Constr Build Mater 26(1):565–573
Limbachiya M, Meddah MS, Ouchagour Y (2012) Use of recycled concrete aggregate in fly-ash concrete. Constr Build Mater 27(1):439–449
Liu H, Liu C, Bai G, Zhu C (2020) Study on the effect of chloride ion ingress on the pore structure of the attached mortar of recycled concrete coarse aggregate. Constr Build Mater 263:120123
Lye CQ, Dhir RK, Ghataora GS, Li H (2016) Creep strain of recycled aggregate concrete. Constr Build Mater 102:244–259
Majhi RK, Nayak AN (2019) Bond, durability and microstructural characteristics of ground granulated blast furnace slag based recycled aggregate concrete. Constr Build Mater 212:578–595
Mao Y, Liu J, Shi C (2021) Autogenous shrinkage and drying shrinkage of recycled aggregate concrete: a review. J Clean Prod 295:126435
Massaad G, Rozière E, Loukili A, Izoret L (2016) Advanced testing and performance specifications for the cementitious materials under external sulfate attacks. Constr Build Mater 127:918–931
Matias D, de Brito J, Rosa A, Pedro D (2014) Durability of concrete with recycled coarse aggregates: influence of superplasticizers. J Mater Civ Eng 26(7):06014011
Mehta PK, Monteiro P (2006) Concrete: microstructure, properties, and materials
Neville AM (1995) Properties of concrete, vol 4. Longman, London, p 1995
Papadakis VG, Vayenas CG, Fardis MN (1991) Fundamental modelling and experimental investigation of concrete carbonation. Mater J 88(4):363–373
Pedro D, De Brito J, Evangelista L (2014) Influence of the use of recycled concrete aggregates from different sources on structural concrete. Constr Build Mater 71:141–151
Qi B, Gao J, Chen F, Shen D (2017) Evaluation of the damage process of recycled aggregate concrete under sulfate attack and wetting-drying cycles. Constr Build Mater 138:254–262
Ridzuan ARM, Ibrahim A, Ismail AMM, Diah ABM. (2005) Durability performance of recycled aggregate concrete. In: Dhir RK, Dyer TD, Newlands MD (eds) International conference on global construction: ultimate concrete opportunities: achieving sustainability in construction. Thomas Telford, London, UK, pp 193–202
Santhanam M, Cohen MD, Olek J (2002) Mechanism of sulfate attack: a fresh look: Part 1: summary of experimental results. Cem Concr Res 32(6):915–921
Sasanipour H, Aslani F, Taherinezhad J (2019) Effect of silica fume on durability of self-compacting concrete made with waste recycled concrete aggregates. Constr Build Mater 227:116598

Seara-Paz S, González-Fonteboa B, Martínez-Abella F, González-Taboada I (2016) Time-dependent behaviour of structural concrete made with recycled coarse aggregates. Creep and shrinkage. Constr Build Mater 122:95–109

Silva RV, De Brito J, Dhir RK (2015a) Comparative analysis of existing prediction models on the creep behaviour of recycled aggregate concrete. Eng Struct 100:31–42

Silva RV, Neves R, De Brito J, Dhir RK (2015b) Carbonation behaviour of recycled aggregate concrete. Cement Concr Compos 62:22–32

Silva RV, de Brito J, Dhir RK (2015c) Prediction of the shrinkage behavior of recycled aggregate concrete: a review. Constr Build Mater 77:327–339

Singh N, Singh SP (2016) Carbonation resistance and microstructural analysis of low and high-volume fly ash self-compacting concrete containing recycled concrete aggregates. Constr Build Mater 127:828–842

Somna R, Jaturapitakkul C, Amde AM (2012) Effect of ground fly ash and ground bagasse ash on the durability of recycled aggregate concrete. Cement Concr Compos 34(7):848–854

Sun H, Wang PM, Sun JY (2006) Study on the gas anti permeability and carbonation resistance of recycled concrete. J Build Mater 9:91–96

Xiao J, Lei B, Zhang C (2012) On carbonation behavior of recycled aggregate concrete. Science China Technol Sci 55:2609–2616

Yildirim ST, Meyer C, Herfellner S (2015) Effects of internal curing on the strength, drying shrinkage and freeze–thaw resistance of concrete containing recycled concrete aggregates. Constr Build Mater 91:288–296

Zhang H, Ji T, Liu H (2020a) Performance evolution of recycled aggregate concrete (RAC) exposed to external sulfate attacks under full-soaking and dry-wet cycling conditions. Constr Build Mater 248:118675

Zhang H, Wang YY, Lehman DE, Geng Y (2020b) Autogenous-shrinkage model for concrete with coarse and fine recycled aggregate. Cement Concr Compos 111:103600

Zhang H, Wang Y, Lehman DE, Geng Y, Kuder K (2020c) Time-dependent drying shrinkage model for concrete with coarse and fine recycled aggregate. Cement Concr Compos 105:103426

Zhang P, Bao J (2023) Durability of high-performance recycled aggregate concrete. In: Multi-functional concrete with recycled aggregates. Woodhead Publishing, pp 85–125

Chapter 11
Recent Advances, Recommendations, and Way Forward

Preview

The use of concrete construction has never been so rampant before and the demand is continuously rising. As a result, the environment is overburdened with the increased CO_2 emissions associated with cement and exploitation of limited natural resources to produce other constituents of concrete such as coarse or fine aggregates. This book has mainly focussed on the recycling and reuse of construction demolition wastes as alternative coarse aggregates and partial cement substitution by other supplementary cementitious materials (SCMs) to produce sustainable concrete. This combination not only helps in managing construction and demolition waste but also enhances the performance and durability of concrete. The inclusion of SCMs has proven to compensate some of the inherent inferiorities associated with coarse RCA based concrete which is discussed previously in other chapters. However, their practical application presents several challenges that need to be addressed to maximize their potential. This chapter explores concerns associated with this recycled concrete and available recent studies to address these issues.

11.1 Challenges of RAC and Recent Advances

Concrete constituted of coarse RCA displays greater variability in quality due to randomness in the source of parent concrete, crushing methods and other impurities present in it. We have already discussed this where higher coefficient of variation in the material properties led to greater partial safety factor in the shear design equation for beams and slabs. Stringent quality control measures can be employed during the collection, sorting, and processing of recycled aggregates to produce uniform aggregates. Advanced processing techniques, such as thermal treatment, mechanical rubbing, and the use of magnetic separators, can help remove impurities and enhance aggregate quality.

S. Sunayana and S. V. Barai, *Green Concrete*, Springer Transactions in Civil and Environmental Engineering, https://doi.org/10.1007/978-981-96-6982-0_11

SCMs has been found to compensate for inferior mechanical properties of RAC. Shi et al. (2018) observed that silica fume and nano-silica used at a cement substitution ratio of 0.63 and 0.13 can increase the compressive strength by 55% and 18% respectively. Similarly, fly ash used as 76% cement replacement can increase compressive strength by 39%. Application of SCMs can affect the setting time, workability, and early-age strength of RAC. Nano-particles, which are finer than 100 nm, has also been applied in RAC to act as filler, and also improve the hydration reaction as well as pozzolanic reaction by providing additional nucleation sites. Nanomaterials which are applied in different studies for RAC include nano-silica (Mukherjee and Barai 2014), nano-alumina (Mohammed et al. 2019), nano-clay (Alhamad 2018), and nano calcium carbonate (Gao et al. 2020) etc. So, it further demands proper mix design to achieve desired properties which can be challenging. The pozzolanic reactions of SCMs can vary significantly based on their type, source, and fineness, complicating the prediction of RAC behaviour. Different types and proportions of SCMs need to be chosen carefully based on specific performance requirements. Standardized testing methods and guidelines can be developed for using SCMs in RAC.

The slump of concrete constituted of coarse RCA has been found to be lower than NAC primarily due to their greater water absorption capacity (Lotfi et al. 2015), rougher surface texture and irregular shape (Verian et al. 2018). 5–15% of extra water may be added to RAC to achieve same workability as that of NAC (Verian et al. 2018). The use of chemical admixtures, such as water reducers, superplasticizers, and shrinkage-reducing agents, can also improve the workability and mechanical properties of RAC.

In comparison to natural aggregates, recycled aggregates have a higher porosity and lower density, which results in a decrease in mechanical properties of concrete. The attached mortar on recycled aggregates leads to the formation of two interfacial transition zone (ITZ) and more weak zones in RAC, further affecting its strength and durability. Optimized mix designs, methods such as particle packing tailored to the specific properties of recycled concrete aggregates and SCMs can help achieve desired performance criteria. Different methods recently developed for modifying the coarse RCA are as follows: (i) pre-soaking coarse RCA in a solution of sodium silicate (Shi et al. 2018) which reacts with CH to generate more C-S–H gel (ii) soaking the aggregates in a polymer solution for a small duration; for example: soaking in for 5 min in silane or siloxane (Spaeth and Tegguer 2013) can improve the old mortar layer of coarse RCA by filling the existing pores and microcracks with the polymer molecules. (iii) the old mortar content reduces by treating it with acid or thermal methods. Strong acids like HCl, H_2SO_4 and HNO_3 can be used for pre-soaking coarse RCA which can improve the compressive strength by maximum up to 35% (Saravanakumar et al. 2016). As compared to the strong acids, mild acetic acid can be more economical as well as effective in reducing water absorption; for example: 1–3% concentrated acid can decrease water absorption by 9–19% (Wang et al. 2017). In the thermal method, the coarse RCA is subjected to heat in order to dehydrate the old mortar so that it loosens and get detached from the aggregate surface (Wang et al. 2021). Microwave technique used for 1 min in RAC can reduce the old mortar thereby reducing the water absorption from 4.2 to 2.8% (Akbarnezhad et al. 2011).

RAC, due to its inherent weakness in the interface regions can be more susceptible to harsh environments posing durability concerns such as chemical attacks, corrosion, and carbonation. Higher porosity of coarse RCA leads to greater possibility of carbonation when used in concrete (Wang et al. 2021). This is evidenced from the fact that the carbonation depth increases by 2.5 times in RAC as compared to NAC. Chloride permeability also increases by two times when RCA is used in concrete (Guo et al. 2018). The properties of RAC can be enhanced through microbial carbonate precipitation using bio deposition with ureolytic bacteria (Singh et al 2018). The technique can be employed to lower the water absorption of RCA, improves compressive strength by 16% and elastic modulus by 27% (Wang et al. 2021).

The long-term durability of this concrete such as creep, shrinkage, freeze–thaw resistance, alkali-silica reaction etc. are essential which must be considered in the design process. The use of fibres, various surface treatment methods, and internal curing can help in mitigating issues related to shrinkage, creep, and cracking. Long-term performance monitoring and field studies can provide valuable data to validate the durability of RAC with SCMs under real-world conditions.

The production process of coarse RCA, besides the inferior mechanical behaviour, serves as one of the obstructions towards promoting wider application of RAC. Other constraints include (i) lack of information about the source of parent concrete (Ma et al. 2020), (ii) unavailability of modern recycling technologies (Wu et al. 2019), (iii) challenges at construction sites, (iv) inadequate market facilities for the supply of recycled materials (Chi et al. 2020), (v) lack of public awareness and enough confidence towards structural safety, and (vi) lack of political will and proper audit from the authorities (Aslam et al. 2020). The key constraints in China, as reported by Ma et al. (2020), include improper management and lack of proper information about the qualities and quantities of CDW, unregulated landfilling at greater cost, economical aspects, and the absence of political will and cooperation.

11.2 Way Forward

To facilitate wider use of coarse RCA and compensate for the inferiorities, various methods are suggested as discussed previously. Bio-deposition method using bacteria, treatment of coarse RCA with acetic acid, use of plant-based fibres such as hemicellulose, lignin etc. (Wang et al. 2021). However, integration of these methods must ensure that the hydration reaction is not disturbed. Addition of nanoparticles is effective for enhancing the mechanical properties of concrete if it is dispersed well in the concrete mix. For this, a cost-effective and faster method will be highly needed. Higher variability of RAC and SCMs properties results in higher partial safety factor in shear to achieve same level of structural reliability. Comprehensive standards and guidelines are needed to be prepared by collaborating with industry stakeholders, regulatory bodies, and research institutions (Kabirifar et al. 2020). This will facilitate the use of RAC and SCMs in practice. Hao et al. (2020) suggested measures such as proper segregation and sorting of wastes, providing benefits and subsidy for recycling activities as well as enhancing the landfill dumping penalty. Region-specific guidelines that consider local materials, practices, and environmental conditions will support consistent and effective implementation. Cost-effective processing and transportation methods can improve the economic viability of using recycled aggregates and SCMs. Centralized recycling facilities and supply chains can be developed which can streamline the logistics and ensure a steady supply of high-quality recycled materials.

References

Akbarnezhad A, Ong KCG, Zhang MH, Tam CT, Foo TWJ (2011) Microwave-assisted beneficiation of recycled concrete aggregates. Constr Build Mater 25(8):3469–3479

Alhamad EG (2018) Influence of Nano-clay on the behavior of recycled aggregate concrete. Cairo University Giza, [M.Sc.] Egypt

Aslam MS, Huang B, Cui L (2020) Review of construction and demolition waste management in China and USA. J Environ Manage 264:110445

Chi B, Lu W, Ye M, Bao Z, Zhang X (2020) Construction waste minimization in green building: A comparative analysis of LEED-NC 2009 certified projects in the US and China. J Clean Prod 256:120749

Gao C, Huang L, Yan L, Jin R, Chen H (2020) Mechanical properties of recycled aggregate concrete modified by nano-particles. Constr Build Mater 241:118030

Guo H, Shi C, Guan X, Zhu J, Ding Y, Ling TC, Zhang H, Wang Y (2018) Durability of recycled aggregate concrete–A review. Cement Concr Compos 89:251–259

Hao J, Yuan H, Liu J, Chin CS, Lu W (2019) A model for assessing the economic performance of construction waste reduction. J Clean Prod 232:427–440

Kabirifar K, Mojtahedi M, Wang C, Tam VW (2020) Construction and demolition waste management contributing factors coupled with reduce, reuse, and recycle strategies for effective waste management: a review. J Clean Prod 263:121265

Lotfi S, Eggimann M, Wagner E, Mróz R, Deja J (2015) Performance of recycled aggregate concrete based on a new concrete recycling technology. Constr Build Mater 95:243–256

Ma M, Tam VW, Le KN, Li W (2020) Challenges in current construction and demolition waste recycling: a China study. Waste Manage 118:610–625

Mohammed RK, Abdel-Hamead AA, Othman FM (2018) Effect of nano-alumina on microstructure and mechanical properties of recycled concrete. J Eng Sustain Dev 22(2):90–103

Mukharjee BB, Barai SV (2014) Influence of incorporation of nano-silica and recycled aggregates on compressive strength and microstructure of concrete. Constr Build Mater 71:570–578

Saravanakumar P, Abhiram K, Manoj B (2016) Properties of treated recycled aggregates and its influence on concrete strength characteristics. Constr Build Mater 111:611–617

Shi C, Wu Z, Cao Z, Ling TC, Zheng J (2018) Performance of mortar prepared with recycled concrete aggregate enhanced by CO2 and pozzolan slurry. Cement Concr Compos 86:130–138

Singh LP, Bisht V, Aswathy MS, Chaurasia L, Gupta S (2018) Studies on performance enhancement of recycled aggregate by incorporating bio and nano materials. Constr Build Mater 181:217–226

Spaeth V, Tegguer AD (2013) Improvement of recycled concrete aggregate properties by polymer treatments. Int J Sustain Built Environ 2(2):143–152

Verian KP, Ashraf W, Cao Y (2018) Properties of recycled concrete aggregate and their influence in new concrete production. Resour Conserv Recycl 133:30–49

Wang L, Wang J, Qian X, Chen P, Xu Y, Guo J (2017) An environmentally friendly method to improve the quality of recycled concrete aggregates. Constr Build Mater 144:432–441

Wang B, Yan L, Fu Q, Kasal B (2021) A comprehensive review on recycled aggregate and recycled aggregate concrete. Resour Conserv Recycl 171:105565

Wu Z, Yu AT, Wang H, Wei Y, Huo X (2019) Driving factors for construction waste minimization: empirical studies in Hong Kong and Shenzhen. J Green Build 14(4):155–167

MIX
Papier aus verantwortungsvollen Quellen
Paper from responsible sources
FSC® C105338

FSC
www.fsc.org

If you have any concerns about our products,
you can contact us on
ProductSafety@springernature.com

In case Publisher is established outside the EU,
the EU authorized representative is:
Springer Nature Customer Service Center GmbH
Europaplatz 3, 69115 Heidelberg, Germany

Printed by Libri Plureos GmbH
in Hamburg, Germany